AF247611

LOW-NOISE MICROWAVE AMPLIFIERS

H. N. DAGLISH   J. G. ARMSTRONG   J. C. WALLING
C. A. P. FOXELL

LOW-NOISE MICROWAVE AMPLIFIERS

CAMBRIDGE AT THE UNIVERSITY PRESS 1968

Published in association with

THE INSTITUTION OF ELECTRICAL ENGINEERS

Published by the Syndics of the Cambridge University Press
Bentley House, 200 Euston Road, London, N.W.1
American Branch: 32 East 57th Street, New York, N.Y.10022

Standard Book Number: 521   07402   9

Printed in Great Britain
at the University Printing House, Cambridge
(Brooke Crutchley, University Printer)

CONTENTS

PREFACE

Low-noise amplifiers of various types are being used in a wide variety
of applications, ranging from the reception of signals from radio
sources in outer space and from communication satellites, to terrestrial communication links and highly sensitive radar receivers.
We feel that a general review of the whole field of low-noise microwave
amplifiers is timely and may enable the important advantages of each
of the many types to be more clearly recognised. We have endeavoured
to describe the properties of each type of amplifier in similar terms
and in a similar sequence, although, with such a wide range of devices,
it has not always been possible to adhere strictly to this pattern.

H. N. DAGLISH
Post Office Research Department
Dollis Hill, London, N.W. 2

J. G. ARMSTRONG
Standard Telephone and Cables Ltd
Paignton, Devon

J. C. WALLING
Mullard Research Laboratories
Redhill, Surrey

C. A. P. FOXELL
Associated Semiconductor Manufacturers Ltd
Hirst Research Centre
Wembley, Middx.

February 1968

ACKNOWLEDGMENTS

The opinions and comments, omissions and errors in this monograph are solely the responsibility of the authors, who gratefully acknowledge, however, the great assistance of their colleagues, both in industry and in Government establishments, in its preparation. In particular, acknowledgments are due to J. S. Lamming and K. Wilson, T. H. Oxley and M. K. McPhun who acted as coauthors of the sections on transistor amplifiers, mixer diodes and tunnel-diode amplifiers respectively and to E. W. Houghton, C. S. Aitchison, J. W. Carter, D. Geden and P. A. Watson whose valuable suggestions have been incorporated in the text.

The preparation of this monograph would have been impossible without making use of the store of information within our respective organisations. We therefore wish to acknowledge with thanks, the permission received from the Directors of Mullard Ltd, Standard Telephones and Cables Ltd, Associated Semiconductor Manufacturers Ltd and the Senior Director, Development, of the Post Office to make use of such information.

Illustrations

The authors are grateful to all those who have permitted the reproduction of illustrations. The illustrations which originate within the authors' organisations are listed below and those from other sources are acknowledged in the appropriate captions.

STC Ltd: Figures 3, 4, 5 and 6 and Plate 1

Mullard Ltd: Figures 9, 11 and 25–35 and Plates 4, 13 and 15–20

Post Office (Crown copyright): Figures 19–21 and 24 and Plates 3, 8 and 10

1 INTRODUCTION

This monograph has been called 'Low-noise microwave amplifiers', but the title might, with equal validity, have been 'Low-noise micro-wave receivers', because the importance of the devices discussed lies primarily in their suitability as preamplifiers in receiving systems, such as those used for radioastronomy, radar, satellite communications and allied purposes.

Low-noise receivers are particularly important for the microwave region of the frequency spectrum because, in this region, full advantage can be taken of a sensitive receiving system. The environment of most radiocommunication systems consists of the sky above and the Earth beneath, and both are sources of noise. The noise radiated into the system by the Earth is determined by Planck's law, which states that the noise power radiated by a body with emissivity ϵ at a temperature $T\,°\mathrm{K}$ in a frequency band Δf about a frequency f is given by

$$P = \frac{\epsilon h f}{\exp\left(hf/kT\right) - 1}\Delta f \tag{1.1}$$

for each propagating mode of the system.

In a microwave system, the properties of the system are normally such as to restrict the number of modes to be taken into consideration to one. In addition, for almost all the systems we are going to discuss, the inequality

$$hf \ll kT \tag{1.2}$$

is applicable. The equation for noise power can then be simplified, for a single-mode system, to

$$P = \epsilon k T \Delta f \tag{1.3}$$

It follows from this equation that the noise from the Earth, which can be regarded as a 'black' body having unit emissivity, amounts to $-114\,\mathrm{dBm}$ per MHz of bandwidth. By the use of directional aerials, it is possible to exclude almost all the Earth noise from the receiving system, except at very low angles of elevation, but noise from the sky is always present.

Instead of expressing this noise in absolute values, it is convenient to use the concept of 'equivalent noise temperature', by rearranging eqn. 1.3. If the noise power due to the sky, or any other source, is P_n in a single mode, and in a bandwidth Δf, then the equivalent noise temperature is given by

$$T_e = \frac{P_n}{k\,\Delta f} \tag{1.4}$$

Noise from the sky, expressed as an equivalent noise temperature, is shown in Fig. 1 for several angles of elevation. At the lower frequencies, the noise is mainly extraterrestrial, originating in the various radio sources of outer space. The level of noise is greater in certain directions in space, and of course, with highly directive aerials, it is possible to resolve some of the radiation into various separate radio stars. For communication purposes this galactic noise sets the ultimate background-noise level of the system. For the radioastronomer, the position is slightly different—the problem is that of detecting and measuring specific noise sources among the general background.

At higher frequencies, the galactic noise rapidly diminishes but noise generated in the Earth's atmosphere becomes important. Water vapour and oxygen both have absorption lines in this region of the spectrum, and therefore the molecules of these substances will not only attenuate signals passing through the atmosphere, but will also add thermal noise in the appropriate frequency bands as indicated in Fig. 1. There will also be contributions from any particles present, such as raindrops, snowflakes, dust, sand etc.

In addition to all these sources, there is a component consisting of radiation flux apparently distributed uniformly throughout space. This radiation, which is equivalent to an additional noise temperature of a few degrees, has been explained theoretically as the residue of the 'fireball' associated with a very early phase of the Universe.

It is clear from Fig. 1 that the 1–20 GHz microwave band is a desirable one for long-distance communication or highly sensitive radar systems, and that the use of a low-noise receiving device is essential to take maximum advantage of this low-noise 'window' in

the frequency spectrum. It is equally clear that radioastronomers who
wish to study phenomena at these frequencies must also use very-low-
noise receivers. We shall therefore be concerned in this monograph
primarily with the 1–20 GHz region, although some of the techniques
are useful at higher and lower frequencies.

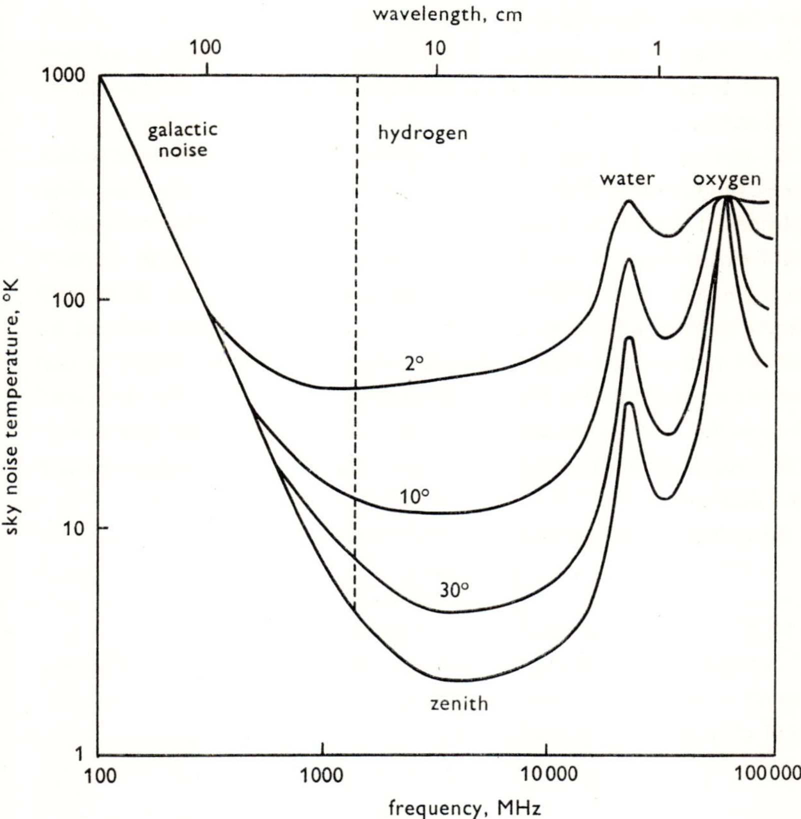

Fig. 1   Noise temperature of the sky at various angles above the horizon

In order to provide a measure of the additional noise introduced by
processes within the receiver, this can be compared with the noise
produced by a 'black body'. A matched termination is used to provide
the microwave equivalent of a black body.

The noise factor† F of a 2-port device at a specified input frequency is the ratio of the total noise power per unit bandwidth (at the appropriate output frequency) available at the output to that portion of this power engendered at the input frequency by an input termination held at a standard reference temperature. The formal definition referred to does not specify the reference temperature. However, although other values have occasionally been used (e.g. $77\,^{\circ}$K), we consider that, to avoid confusion, only one reference temperature should be used and all noise factors quoted in this monograph are based on $290\,^{\circ}$K as the reference temperature.

The concept of noise factor is useful for traditional microwave devices, but for some of the newer devices, such as masers or parametric amplifiers, it gives numerical values which differ very little from unity, and the *excess* noise factor $(F-1)$ is more significant. However, a more useful concept is the effective input noise temperature T_n (often referred to as amplifier noise temperature), which is also defined in terms of a matched input termination. The effective input noise temperature is defined as the temperature at which the input termination must be held to produce an output noise power, per unit bandwidth, double that which would occur if the termination were cooled to absolute zero.

The relationship between the two parameters is

$$T_n = T^*(F-1) \tag{1.5}$$

where $T^* = 290\,^{\circ}$K, the reference temperature.

If several devices are connected in tandem, the overall effective input noise temperature can be calculated, provided that the available gain of each device is known, from

$$T_n = T_1 + \frac{1}{G_1}\left\{T_2 + \frac{1}{G_2}(T_3 + \ldots \right. \tag{1.6}$$

where the subscripts refer to the first, second and third devices.

Hence, if the gain of the first stage is low (say 10–15 dB), in many applications the noise performance of the second stage is also important. Some of the devices described in this monograph are equally

---

† Alternatively known as noise figure, but noise factor is the preferred term according to BS 204:1960

4

important for separate use as moderately-low-noise receivers or as a second stage following one of the other very-low-noise amplifiers.

The performance of a complete system can also be described in terms of an effective noise temperature. (The reference point for this is usually quoted as the input port of the first-stage amplifier.) As well as external noise from the sky and the Earth etc. and internal noise from the receiver, there is also a contribution from thermal noise caused by the losses in the interconnections between the aerial and the amplifier. An attenuation, in this interconnection, of 0·1 dB will not only reduce the level of signal (and external noise) reaching the amplifier, but will add 7 °K to the system noise temperature. (It is assumed that the interconnections are at about 290 °K.) This additional noise may be very important in some applications, and may influence the design of the receiving equipment, since this may have to be mounted very close to the aerial, to reduce the length of the interconnecting transmission line.

Fig. 2 shows the typical ranges of effective input noise temperature for various amplifiers. The amplifiers include the following: travelling-wave tubes and other electron-beam devices, varactor-diode parametric amplifiers, tunnel-diode amplifiers, masers and microwave transistor amplifiers. These various devices all have effective input noise temperature less than about 1000 °K, as shown in Fig. 2; the values do not include any noise contributed by any following stage or stages.

In a microwave system, an amplifier must have separate input and output ports. The input at least must normally be well matched, and the isolation of the output from the input must be good. These conditions are met in different ways in the different devices discussed in this monograph. The transistor amplifier and the electron-beam travelling-wave amplifiers are basically 2-port devices, although, in the latter, adequate precautions must be taken to attenuate any backward waves. The travelling-wave maser in its simplest form provides gain in only one direction but would introduce only a moderate reverse attenuation. Such an amplifier would be prone to parasitic oscillation unless considerable care were taken to maintain a well matched system, both inside and outside the amplifier. Additional ferrite material is therefore incorporated to ensure a high reverse loss under all circumstances.

The remaining amplifiers—tunnel-diode, varactor-diode and earlier forms of cavity-maser amplifiers—are reflection rather than transmission devices, i.e. they can be regarded as single-port elements which present a negative impedance to the remainder of the circuit. There is

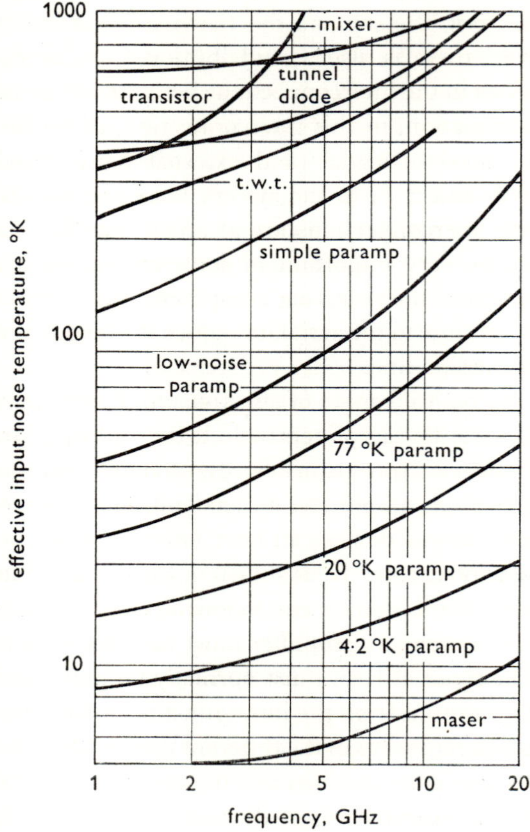

Fig. 2   Typical ranges of effective input noise temperature for low-noise amplifiers (all contributions from following stages ignored)

a considerable body of literature devoted to the properties of resonant circuits containing one or more active elements, particularly when coupled to passive elements to control the gain–bandwidth properties of the device. In order to incorporate a negative-impedance element

6

into a microwave circuit it is necessary to separate the input from the output signal. Although hybrid junctions and directional couplers are sometimes used, they degrade the potential performance, and the normal method is to use an additional circuit element, a circulator, in which nonreciprocal directional properties are introduced by using ferrite material under the influence of a magnetic field.

Circulators are available in many configurations, both in stripline and waveguide, with either three ports or four ports, and, on combining several units, devices with five or more ports are made available. The impedance characteristics of the circulator inevitably affect the performance of the associated amplifier, and some attempt has been made recently to control the circulator characteristics to match the associated amplifier. For the best possible amplifier performance the use of separate components must be abandoned and the active element, tuning circuits, and multiport circulator integrated into a single unit. In this way, the amplifier designer is free to specify the impedance characteristics within the device without the complication inevitably introduced when individual components have to be mated using standard 50 or 75 Ω connectors. Some tunnel-diode amplifiers and a few parametric amplifiers have already been designed on this basis.

Although the various devices mentioned above have in common the property of a relatively low level of internal noise, many of their other properties are also of importance, depending on the particular application. With the growing use of satellite communications, wider bandwidths are becoming necessary without sacrifice of noise performance or the ability to operate in the presence of a high-power transmitter. Linearity is important, because of the need to minimise intermodulation distortion when multiple carriers are being received. For other purposes, a relatively narrow instantaneous bandwidth is acceptable, or even preferable, provided that the device can be readily tuned electrically over a very wide range. In radar and in some communications applications, high signal levels may be present at times, and the amplifier must be capable of withstanding these without damage, with a rapid recovery of normal sensitivity when the high-level signal is removed.

These and the other important properties of the various devices are discussed in detail in the following chapters, together with some indication of the performances which are likely to be achieved in the near future, as a result of continuing development. In order to complete the picture, recent developments in detector and mixer diodes have been included, because of their importance as input devices or as second stages following the other low-noise amplifiers. It is not, however, our intention to describe the design and construction of mixers and detector mounts, but to concentrate on the improvements in performance of such devices which have been made possible by recent advances in the technology of the appropriate diodes.

For the design and optimum use of low-noise devices, it is necessary to solve many problems in the accurate measurement of noise temperatures. It is felt, however, that a discussion of these problems would be beyond the scope of the present monograph.

2 TRAVELLING-WAVE TUBES

2.1 Introduction

The travelling-wave tube (t.w.t.) operates by employing an extended interaction between a moving electron beam and an electromagnetic wave travelling with the beam. The principal features of such a tube are an electron gun which produces the beam of electrons, a slow-wave structure on which the electromagnetic wave is propagated and a focusing system which constrains the electron beam to travel along (and normally within) the slow-wave structure.

A number of possible slow-wave structures exist, but the one most widely used is the helix, as it is simple to construct and offers a very broad bandwidth. A simple diagram of a helix-type t.w.t. is given in Fig. 3. In the example shown, the r.f. wave is coupled to the helix by antennas which may lie across a section of waveguide or inside a coaxial cavity. The r.f. wave travels around the helix at essentially the velocity of light, giving rise to a travelling wave on the axis which moves at some fraction of the velocity of light (typically one-tenth to one-fortieth), determined by the pitch and circumference of the helix. By the application of a suitable voltage to the helix, the electron beam is accelerated to about the same velocity as the slow r.f. wave, when interaction can occur. In this situation, some electrons are retarded while others are accelerated by the r.f. fields. Amplification of the r.f. wave occurs when more electrons are retarded than are accelerated, the energy for amplification coming from the average loss of kinetic energy of the electrons.

In Fig. 3, a region of attenuation is shown midway along the helix. This provides the necessary isolation between input and output to ensure stability. A typical t.w.t. has a gain of 30–40 dB, and reflections at the output and input couplers would cause oscillation if this isolation were not provided. The attenuation naturally removes the forward growing wave from the helix as well as the unwanted reflected wave, but the signal information is carried through the attenuator region by the electron-beam modulation and is re-established on the helix after the attenuator.

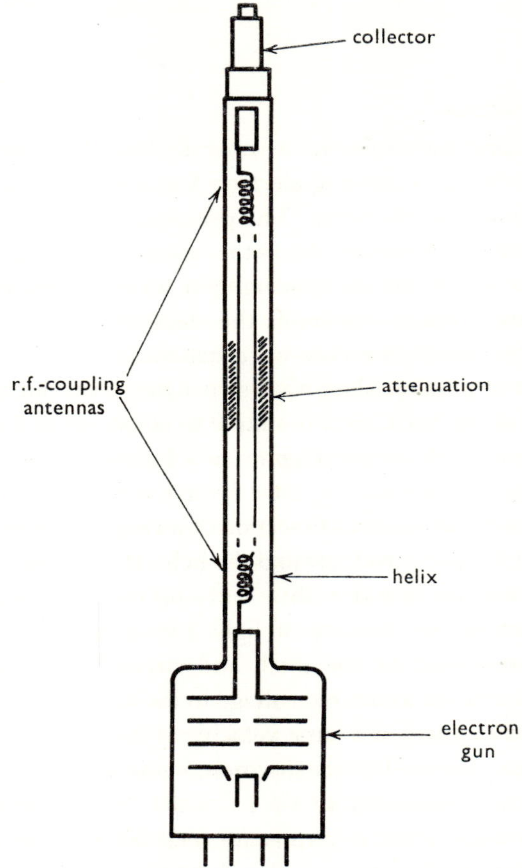

Fig. 3   Schematic of helix-type travelling-wave tube
(external magnet omitted)

A detailed theory of the travelling-wave tube is given by Pierce (1950). It is shown that any disturbance of the slow-wave circuit and electron-beam system gives rise to four waves, of which one is a backward wave and three are forward waves. Of the three forward waves, only one is amplified, and this is the wave used in the operation

of the travelling-wave tube. For small-signal operation, the gain of
a t.w.t. can be expressed as follows:

$$G = A + BCN - \alpha L_1 - L_2 \tag{2.1}$$

In this equation, A represents the initial loss in setting up the growing
wave, and this is typically between -6 and $-10\,\text{dB}$. BCN is the
magnitude of the electronic gain ignoring losses. C is the gain parameter
and is given by

$$C = \left(\frac{KI_0}{4V_0}\right)^{\frac{1}{3}} \tag{2.2}$$

where K represents the interaction impedance between the beam and
circuit. I_0 and V_0 are the current and voltage of the beam.

B is the increasing-wave parameter and is a numerical factor
dependent on the space charge present. The quantity BC is then the
gain in decibels per wavelength, and this is multiplied by the number N
of wavelengths of interaction to give the total electronic gain, ignoring
losses.

L_1 is the cold loss of the helix ignoring the attenuation region, i.e.
the loss of the helix in the absence of the electron beam, and α is the
fraction of that loss which subtracts from the gain. α is less than 1,
usually about $0\cdot4$, and the reason for this is easily seen if the signal is
considered to be carried partly on the helix and partly on the beam;
the part on the beam is not subject to the helix loss.

L_2 is the loss associated with the attenuation region. This region can
be considered as a drift space in which the helix wave is missing, and
the only information carried through it is that contained in the
bunching of the beam. A loss is clearly associated with the removal of
the helix wave, and a further loss is determined by the length of the
drift region as space-charge debunching will occur to some extent in
that region.

The above discussion relates to small-signal operation of a t.w.t.,
but nonlinear effects naturally occur as the output power approaches
saturation. Normally, saturation occurs when the r.f. output is
between 5% and 40% of the d.c. beam power, and at saturation the
gain is several decibels less than the small-signal value.

It was mentioned earlier that the helix slow-wave structure offers a very broad bandwidth because the phase velocity varies very little with frequency. A typical helix t.w.t. has an instantaneous bandwidth of an octave, and at least one tube is available commercially with a bandwidth of two octaves over which the gain variation is less than 6 dB. Higher-impedance, and therefore more efficient, structures such as coupled cavities are used on high-power t.w.t.s, but these have greater variations of phase velocity with frequency and therefore operate over much narrower instantaneous bandwidths.

2.2   Low-noise t.w.t.s

The principal advantages of t.w.t.s over other amplifiers for low-noise applications are very broad bandwidth, relatively high saturation output power giving a wide dynamic range, simplicity of operation and resistance to damage caused by large input signals. Until fairly recently, these advantages applied at frequencies as low as 500 MHz, but improvements in the performance of solid-state devices are steadily increasing the frequency at which t.w.t.s have a clear-cut advantage over other amplifiers.

Disadvantages of the t.w.t. include a comparatively short life and, usually, a fairly large size and weight, although significant progress is being made in both of these areas. The minimum noise temperature that can be achieved is much higher than is possible with, for example, masers and cooled-varactor-diode parametric amplifiers, and this makes the t.w.t. unsuitable for applications calling for the best possible noise performance regardless of bandwidth and dynamic range.

In the following sections, the performance and design parameters of low-noise t.w.t.s are discussed in detail to indicate the present limitations in the design and development of these devices.

2.3   Theoretical noise performance

Any analysis of noise behaviour involves the use of statistical techniques, and an excellent discussion of their application to noise propagation on electron beams is given by Haus (1959).

Where only linear lossless processes are involved, it is shown that two invariants exist, termed S and Π. Their relationship to the statistical description of the noise is

$$S = [\Phi\Psi - \{I_m(\Theta)\}^2]^{\frac{1}{2}} \qquad (2.3)$$

$$\Pi = \mathrm{Re}\,(\Theta) \qquad (2.4)$$

where

Φ = selfpower density spectrum of kinetic noise-voltage variations

Ψ = selfpower density spectrum of noise-current variations

Θ = cross-power density spectrum between kinetic noise-voltage and noise-current variations.

For a lossless amplifier structure it can be shown that the minimum noise factor F_{min} has the form

$$F_{min} = 1 + \left(1 - \frac{1}{G}\right) \frac{2\pi}{kT}\ (S - \Pi) \qquad (2.5)$$

where $\qquad\qquad G$ = amplifier gain

and $\qquad\qquad T$ = input circuit temperature

If we consider a low-noise t.w.t. having full shot noise and Rack velocity noise at the cathode, and having no nonlinear processes present, Θ is zero. In this case, S is given by

$$S = \left(1 - \frac{\pi}{4}\right)^{\frac{1}{2}} \frac{kT_c}{\pi} \qquad (2.6)$$

where T_c is the cathode temperature. For a high-gain amplifier with $T = 300\,^{\circ}\mathrm{K}$ and $T_c = 1000\,^{\circ}\mathrm{K}$, from eqns. 2.6 and 2.5, a minimum noise factor of about 6 dB is obtained.

In practice, nonlinear processes occur when the beam velocity is very low, i.e. comparable to the thermal electron velocities, and under these conditions S and Π are not invariant. What happens is that the amount of noise, crudely measured by S, can be reduced, and correlation, measured by Π, can be set up. In this way the quantity $(S - \Pi)$ can be reduced, resulting in a lower optimum noise factor. Practical application of this is mentioned in the next section.

2.4 Electron gun

As far as noise performance is concerned, the most vital section of the tube is the cathode region of the electron gun, as the processes occurring in this region dictate the inherent beam noisiness. Subsequent processes can degrade the noise performance obtainable using a given electron beam, but no reduction in slow-wave noise is possible once the beam has been accelerated to a velocity at which its behaviour is linear.

The emission processes of the cathode give rise to noise in the form of random current variations (shot noise) and random voltage variations (Rack velocity noise) from all parts of the active surface, and spatial variations in emission from different parts of the surface. To minimise noise it is important to use as low a temperature as possible to limit velocity variations, while ensuring sufficient emission to give space-charge-limited flow to reduce the current variations. To limit variations in different parts of the beam, it is essential to have a very uniform emitting surface.

At present, oxide coatings are the most suitable emitting surfaces, and these are carefully prepared to give a smooth uniform coating. Usually, the surface roughness is less than $3\,\mu$m peak–trough. It is possible that in the future 'cold' cathodes, such as the tunnel cathode, may be developed for use in low-noise t.w.t.s. If this were to happen, the reduction in temperature from about $1000\,^{\circ}$K for an oxide cathode could result in a much reduced beam noise.

The other part of the cathode region is the low-velocity drift region which extends to the point where the beam velocity is significantly larger than the random thermal velocities of the electrons. Apart from classical space-charge smoothing, it has been firmly established, initially by Currie (1958), that further reduction of beam noisiness is possible when the low-velocity drift region is extended by using a highly divergent electric field at the cathode, with a strong axial magnetic field present to impose radial constraint on the beam. It has not yet been possible for a full theoretical analysis of the nonlinear processes to be made, but the limited work that has been carried out suggests that the noise power can be reduced, and in addition that correlation can be set up between current and velocity fluctuations to reduce the effective noisiness of the beam. This was mentioned in the previous section discussing the theoretical noise performance.

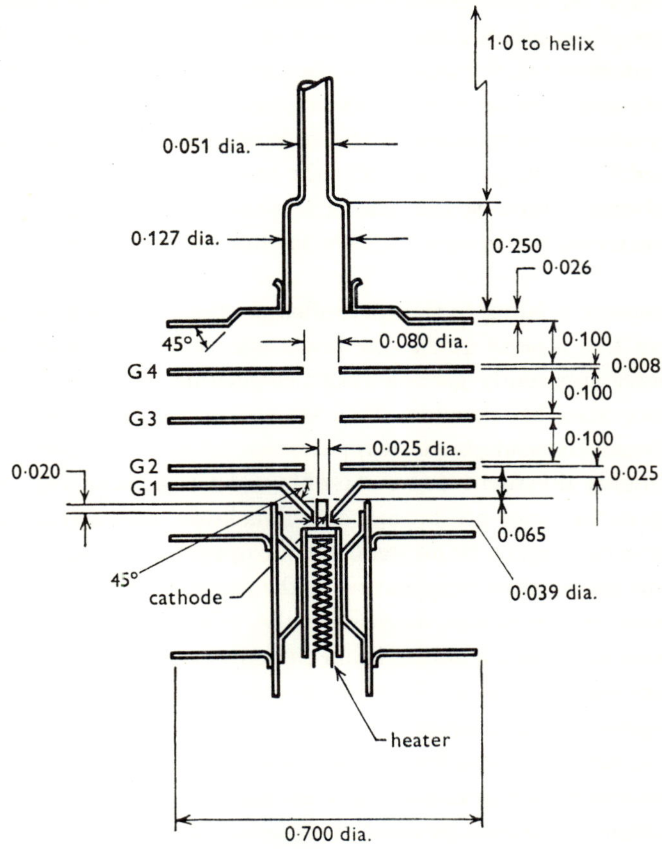

Fig. 4   Typical low-noise electron gun

Dimensions in inches

V drift $= 190$ V $V_{G2} = 4$ V
$V_{G4} = 30$ V $V_{G1} = 3$ V
$V_{G3} = 15$ V

A typical very-low-noise electron gun (Fig. 4) has a positive beam-forming electrode cupped around the cathode pip to produce the divergent electric field. The geometry of this electrode and of the cathode pip are critical for optimum performance, as is the beam space charge, since all of these factors affect the potential profile followed by the beam. The best results have been obtained using a cathode with a sharp edge; with this geometry the electric field is greatly enhanced at the edge of the cathode and an essentially annular beam is produced. The uniformity of the coating at the cathode edge is extremely important, and, although good results have been obtained with the coating sharply terminated at the edge of the cathode pip (using a trimmed film coating, for example), the best results have been obtained by deliberately allowing the coating to extend over the edge of the pip, and this technique is used by most manufacturers of very-low-noise t.w.t.s.

A measure relating to the space-charge effects which is sometimes used for this type of gun is the linear edge-current density, as this parameter apparently gives a closer correlation with observed performance than, for example, the average area current density. From experimental measurements, it appears that the best noise performance with a highly divergent-flow gun is obtained with a linear edge-current density in the range 0·1–0·5 mA/cm. This limits the current that can be used with a given beam diameter for optimum performance, and therefore has an important bearing on the variation of noise factor with saturation output power, discussed later. It has been stated that the best possible values, ignoring beam current, have been obtained with the sharp-edged cathode. However, it is possible that, when a higher than optimum beam current is dictated by other tube-design considerations, a different cathode-edge geometry might improve the noise performance at the higher current. This might be achieved by spreading the current over a broader annulus to reduce the space charge, i.e. to make an average area current density more applicable than the linear edge-current density.

The final factor to be considered in the cathode region is the magnitude of magnetic field, since the noise performance improves steadily with magnetic field for the type of gun being discussed. Armstrong (1963) has shown that, for an uncooled S band tube with a noise factor of 3·4 dB at a cathode magnetic field of 600 Oe, the noise factor improved steadily to 2·4 dB at 2000 Oe. Hammer and Thomas (1964),

using a tube with comparable performance under the same conditions,
obtained a best noise factor of 1 dB using a magnetic field of 4500 Oe
and with the helix maintained at liquid-nitrogen temperature. Under

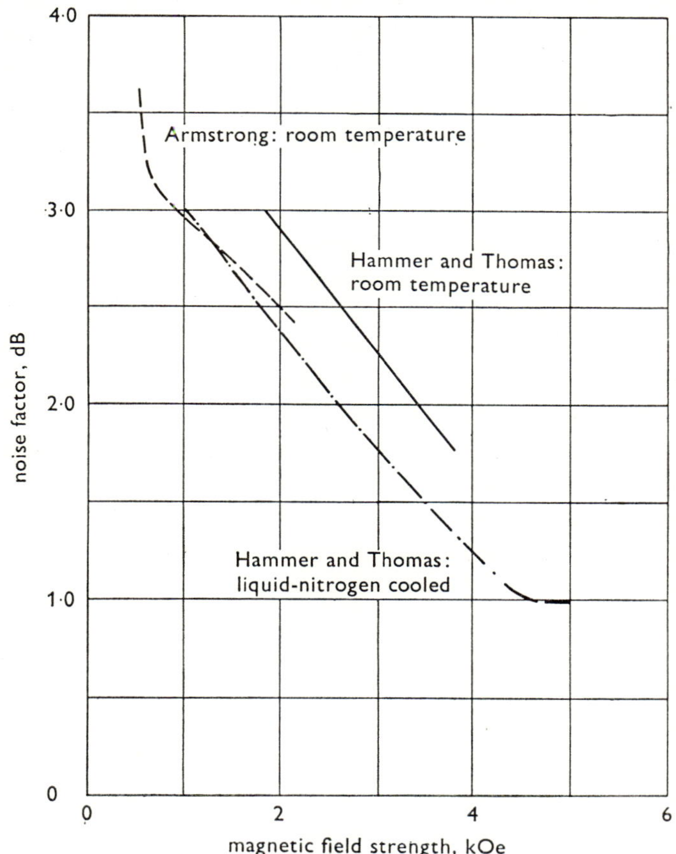

Fig. 5   Noise factor of travelling-wave tube as a function of magnetic field (Hammer
and Thomas data by courtesy of the Institute of Electrical & Electronics
Engineers)

these conditions, the equivalent noise temperature of the electron
beam was estimated to be 51 °K (allowance having been made for
coupler losses etc.), which is quite remarkable in view of the cathode
temperature of 1000 °K. Both of these results are plotted in Fig. 5.

The improved noise performance at elevated magnetic fields is probably associated with enhanced nonlinear interactions in the low-velocity drift region, caused by the tighter confinement of the beam.

The other regions of a low-noise electron gun are the transformer section, where the beam is accelerated to helix potential, and the drift space prior to the start of the helix. In the transformer, the potential profile is controlled by a number of electrodes to optimise the standing-wave ratio of the noise space-charge waves for the particular helix being used, and the drift space is of a length to optimise the phase of the noise standing wave at the helix-entry position. No reduction of beam noisiness occurs in these regions, as the beam processes are linear, but the beam conditions are set to ensure the best performance with the noise that is present.

The design of the transformer involves a compromise, for it must be long enough to ensure flexibility, but an increase in beam noisiness has been observed when the total gun length is large. This increase is thought to be due to the growth with distance of radial-space-charge modes which can interact with the helix to increase the tube noise factor. In tube design, this implies that the lowest practicable helix voltage should be chosen to minimise the impedance transformation required.

2.5 Helix and electron beam

The helix must be of very uniform construction and have a very low loss, at least in the input section, as any loss of signal before it has grown well above the noise level degrades the performance.

For the best noise performance, the helix voltage should be as low as possible to minimise the beam impedance transformation, but, for a comparable design, the optimum helix pitch and diameter are reduced as the voltage is lowered. One limitation on the lowering of the voltage is therefore the difficulty of manufacture and assembly of fine-pitch helices with suitable uniformity and low loss. Another limitation is the difficulty of maintaining beam-current and cathode uniformity. The total current for optimum performance is limited by the requirement for a low linear edge-current density, and the smaller the helix diameter, the smaller the cathode diameter and the lower the optimum beam current.

It is not possible, at present, to use beam convergence and a larger cathode, as electrostatic convergence is ruled out by the need for a divergent electric field at the cathode, and magnetic convergence is impracticable. A given design must use a high cathode magnetic field for the best noise factor, rather than a low cathode field with a high helix field.

One way in which the cathode and helix dimensions can be increased is by designing for a large value of γa, which is a parameter relating to the strength of the r.f. interaction. It is defined as $2\pi a/\lambda_g$, where a is the helix radius and λ_g is the guide wavelength. With a solid-beam t.w.t. it is normal to design for a midband γa of about $1\cdot5$, as this leads to the flattest broadband gain response, but, when a hollow beam is used, midband γa values of about 3 can be used, and in fact a flatter gain response can then be achieved. A low-noise gun using a divergent electric field at the cathode produces a largely hollow beam, and if necessary an annular cathode can accentuate this; so the use of a large γa is practicable. A price has to be paid, however, namely a lower gain per unit length with increasing γa, and this is not always acceptable.

For the best possible noise performance, then, comparatively low voltages and currents must be used and consequently ultra-low-noise t.w.t.s have a limited saturation output power. In general, also, the noise performance of t.w.t.s deteriorates with increasing frequency. This may be associated with fundamental beam limitations, but it is also partly because variations from cathode and helix uniformity become larger in relation to the operating wavelength, signal losses in the input section of the helix increase with frequency, and a larger helix voltage is dictated to keep the helix and the cathode to practicable sizes.

Table 1 gives details of a few of the important results announced in the last few years. Solenoid focusing was used in all of these experiments, and it is seen that, with a magnetic field of about 5000 Oe and with the helix cooled to liquid-nitrogen temperatures, best noise factors ranging from less than 1 dB at u.h.f. to less than 3 dB at X band frequencies are feasible in the laboratory. The corresponding figures for operation at more practicable magnetic fields of 1000 Oe or less, and with no helix refrigeration, are about 1·5 dB at u.h.f. and 4–5 dB at X band frequencies. Not a great deal of work has been done at frequencies above X band, but one American manufacturer produces tubes which can give noise factors of less than 9 dB at 12–18 GHz and less than 13 dB at 26–40 GHz. Saturation output power for the ultra-low-noise tubes is usually in the range − 10 to 0 dBm, although it is somewhat higher for the tubes operating above X band frequencies.

Table 1   *Details of some important low-noise t.w.t. results*

Reference	Israelsen and Peter (1962)	Vehn and Peter (1963)	Armstrong (1963)	Hammer and Thomas (1964)	Nelson and Israelsen (1965)
Frequency, GHz	1·65	0·355	3·0	2·6	12·0
Noise factor, dB	1·7	1·5	2·4	1·0	2·7
Helix voltage, V	135	50	195	390	400
Beam current, μA	55	300	40	107	100
γa	1·7	2·8	1·5	—	3·7
Magnetic field, Oe	4500	940	2000	4500	5000
Helix temperature, °K	77	290	290	77	77

The above discussion and the results mentioned all relate to tubes designed specifically for optimum noise performance. In most practical applications, however, other considerations such as gain, size and weight and saturation output power (which determines dynamic range) cause compromises to be made with respect to noise performance. For a given type of focus mount it is obvious that gain, size and weight are interrelated, but for a given gain the choice of focus mount can greatly affect the size, weight and the noise performance. This is discussed in the next section. As far as the tube design is concerned, where high gain per unit length is an important consideration, it is necessary either to reduce the beam voltage or to increase the current. The first method is generally unattractive, as for optimum noise performance the voltage will already have been chosen as low as physical dimensions readily allow, and a reduction in saturation output power would be incurred with this solution. An increase in beam current can be used with no dimensional problems, and although the noise performance deteriorates, this is compensated for to some extent by an increase in power.

When a high saturation output power is required, it is generally found best to increase both the current and the voltage to obtain the best compromise. If only the voltage is increased, the gain per unit length decreases rapidly and the tube becomes prohibitively long for a given gain, whereas if the current alone is increased the cathode loading becomes excessive and both the noise factor and the tube life deteriorate. When both voltage and current are increased in appropriate proportions, the tube length remains reasonable, and the cathode loading does not become excessive since the higher voltage leads to larger diameters of helix and cathode. However, the noise factor deteriorates as a result of the increases in current and voltage and it is possible that a fundamental limitation applies. Because much larger linear edge-current densities are involved than the optimum for ultra-low-noise guns, space-charge effects are more pronounced and it is not possible for much of the beam to be subjected to the 'ideal' potential profile through the low-velocity-drift region.

In fact, with high-power low-noise tubes, the best noise performance is usually obtained with comparatively slight divergence of electric field at the cathode, and therefore no extended low-velocity region is involved. The best noise factor reported with a high-output-power t.w.t. is 5·5 dB at S band frequencies for an experimental tube capable of 1 W output with a low-level gain of 30 dB. This result was obtained in a solenoid with a field of about 1000 Oe (Raub and Berger, 1963). As an extended low-velocity-drift region does not play a significant part in the performance of these high-power low-noise tubes, it is possible that further research into cathode and electrode geometry could lead to lower noise factors. Such an improvement would depend on developing a design in which nonlinear noise-reduction processes can be made to work on most of the beam.

There is one technique available for increasing the saturation output power and still maintaining optimum conditions in the gun and helix input sections; this is to increase the beam power in the output section by raising the voltage. Ideally, to make full use of this technique, the diameter of the helix and beam should be increased in the higher-voltage output section to maintain an optimum range of γa across the working frequency band. This would obviously introduce mechanical problems in the helix structure, and the beam would have to be expanded, probably by the use of a divergent magnetic field, but the technique is worth considering where it is essential to obtain the lowest possible noise factor coupled with maximum dynamic range.

2.6 Focus mounts

For practical application of low-noise t.w.t.s, the type of focus mount is of major importance. It has already been indicated that the lowest noise factors are obtained by using high-field solenoid mounts, but permanent-magnet focusing is now becoming essential to provide devices suitable for field use. It is also important to minimise size and weight, as the ability to place the amplifier close to the receiving aerial can be valuable in terms of overall system noise factor.

Three types of permanent-magnet focusing are possible: straight-field (s.f.p.m.) systems which give similar magnetic-field profiles to those used in solenoids (but with severe practical limitations on the magnitude of field that can be achieved), reversed-field systems (s.r.p.m.) in which two or more straight-field sections are used and periodic systems (p.p.m.). The magnetic profiles of these three types are depicted in Fig. 6.

The simplest form of s.f.p.m. mount uses a cylindrical steel magnet (or magnet array), with the field focused along the axis of the cylinder. Usually the magnet is shaped and used with suitable polepieces to provide a uniform field over most of the tube length, with an increase in field of up to 50% over the electron gun and, in particular, the cathode region. The main limitation with axially magnetised s.f.p.m. mounts is due to the large external leakage fields which waste most of the flux available from the magnets. To keep the magnet cross-section to a reasonable size it is necessary to use steel magnets which work at much higher flux densities than ferrite magnets, but the average field strengths that can be efficiently achieved are limited to about 1000 Oe with high-coercive-force steel magnets such as Alnico 8. Higher field

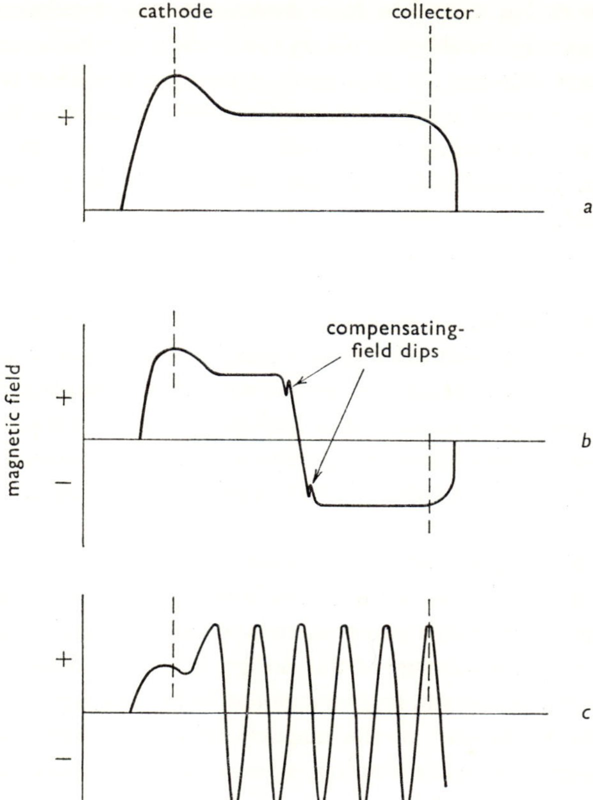

Fig. 6   Typical magnetic-field profiles for focusing low-noise
travelling-wave tubes

    a   Straight field
    b   Reversed field
    c   Periodic field

strengths could be achieved using Platinax, but the high cost of this material virtually limits its use to space applications.

A problem encountered with permanent magnets is the presence of fairly strong transverse fields, particularly when high-energy-product materials are being used. These transverse fields have typical strengths of 10–20 Oe and have to be reduced to about 1 Oe to ensure the best focusing and noise performance with a low-voltage beam. Considerable work has therefore been devoted to the development of field straighteners suitable for use in low-noise-t.w.t. focus mounts. One approach is to use an axial array of thin annular discs of high-permeability material, and an alternative technique is to use helically wound high-permeability wire, which forms a thin robust field straightener. Dimensional details are chosen to limit the effect on the axial field.

The best examples of low-noise t.w.t.s using axially magnetised straight-field focus mounts are provided by the Watkins–Johnson series of packaged ultra-low-noise amplifiers. These cover the frequency range 0·5–18 GHz, with noise factors ranging from about 4 to 9 dB. The gain is about 28 dB and the saturation output power ranges from − 5 dBm at the low frequencies to 0 dBm at the high frequencies. The mounts are magnetically screened, weigh less than 17 lb each, measure 12 in long by 4·5 in diameter, and contain integral power-supply units.

An alternative form of straight-field mount uses radially magnetised magnets, a system in which the distribution of magnets is chosen to give the required magnetic potential variation along the axis (Meyerer, 1964). The major advantage of this design is that the outside of the mount is kept at a uniform magnetic potential by a high-permeability material that forms part of the magnetic circuit. In principle, the magnetic assembly is self-screening and much less total flux is required from the magnets. It is therefore possible to use ferrite magnets which work at much higher field strengths than steel magnets, and the total radial length of active magnet can be much less than the axial length of the air gap. No production low-noise t.w.t.s use this design yet, but Armstrong and his colleagues are developing such a mount for a tube which operates at frequencies from 2 to 4·5 GHz. This tube has a gain of 40 dB and a saturation power of about 18 dBm; midband noise factors as low as 5·5 dB in the p.m. mount are currently being obtained, which compare with best values of about 5 dB in a solenoid. In the same mount, a noise factor of 3·7 dB has been obtained with an ultra-low-noise S band tube. Such a tube and mount are illustrated in Plate 1.

The second type of focusing to be considered is the reversed-field system. In this, two or more straight-field magnets are used with alternate polarities. In this way the leakage flux is reduced considerably even when axially magnetised sections are used. Because an infinitely sharp field reversal cannot be achieved, perturbation of the beam occurs at the reversal and much work has been carried out to produce field profiles close to the reversal region which compensate for this perturbation. Thus additional perturbations are deliberately induced on the electron beam which are approximately equal in magnitude but opposite in phase to that caused by the reversal itself.

The reasoning supporting the use of reversed-field focusing for low-noise t.w.t.s (or the use of combined straight-field and periodic-field focusing) is that beam perturbations, which might in themselves be noisy, should not affect the tube noise factor provided that they occur after significant gain has taken place. Thus the first straight-field section is made to extend over part of the input helix section, and, at beam voltages greater than about 500, noise performance essentially equal to that in a solenoid working at the same average magnetic field strength can be obtained.

At lower beam voltages it becomes more difficult to achieve good focusing through a reversal since the beam perturbation gets pro-gressively larger. In addition, the amount of axial beam energy trans-ferred into rotational energy by the change in flux linking the beam significantly desynchronises the beam and helix waves. Armstrong and his colleagues have studied this problem in detail with a view to putting an ultra-low-noise S band tube in an s.r.p.m. mount. Although consistently good focusing (99·75% transmission at the working voltage of 190) was obtained, and the beam and helix waves were resynchronised by the use of a higher voltage on the output helix section, the best noise factor obtained was 5·7 dB. This was with a tube design which gives less than 4 dB in a solenoid or a straight-field p.m. focus mount. In addition to the relatively high noise factor the gain was extremely critical, and it was concluded that reversed-field focusing is not at present practicable for very-low-voltage tubes.

The third focusing system to be considered is that using periodic permanent magnets (p.p.m.). This offers the lightest weight of all, but again it is limited to fairly high voltages (usually more than 500 V).

Considerable scalloping of the beam occurs with p.p.m. focusing, and if the periodic stack starts before significant gain has taken place, then the noise factor is degraded. A US firm offers a range of p.p.m.-focused low-noise tubes for frequencies of 2–38 GHz, weighing 6–12 lb and with noise factors of 10–16 dB. These tubes have saturation output powers of 10–7 dBm. P.P.M.-focused high-power low-noise tubes with maximum noise factors of 15 and 22 dB at S and X band frequencies are also available, and these tubes have saturation powers of 27 and 33 dBm respectively. To minimise the degradation of noise performance, some manufacturers have used a combined straight-field–p.p.m. system, and, for example, X band packaged tubes weighing only 5·5 lb with noise factors below 8 dB are available. In their development work, General Electric in the USA have made use of radially magnetised steel magnets and have demonstrated that these offer advantages for certain applications.

2.7 R.F. coupling

Whichever method of providing the magnetic field is chosen, the final design is influenced considerably by the choice of the method of coupling the r.f. signal into and out of the tube. The methods available are waveguide, coaxial cavity, coupled helixes and direct connection.

For bandwidths of less than an octave, waveguide coupling is very straightforward in solenoid-focus mounts, where it is easy to maintain an acceptable magnetic-field shape across the waveguide gap, but if the frequency is low and the waveguide is therefore large, it is difficult to bridge the gap in a p.m. mount. Because of this, the use of normal-width waveguide is only attractive at X band and higher frequencies. Waveguide can, however, be reduced in size by dielectric loading. For example, in an S band experiment conducted by Armstrong and his colleagues, the waveguide was loaded with alumina which allowed a reduction in the dimensions to those of normal X band guide, but the bandwidth was limited to about 600 MHz for a v.s.w.r. of less than 2:1. In addition, losses introduced by the dielectric loading amounted to about 0·2 dB for each terminal.

Coaxial cavities are suitable for bandwidths of an octave or possibly a little more, but they also present dimensional problems as the frequency is lowered. This is not as serious as for waveguide coupling because the cavity is contained within the magnet structure, so that no gaps in the magnet have to be bridged, but a restriction is imposed on the minimum internal diameter of the magnets. Because of this, cavity coupling only becomes really attractive at C band and higher frequencies.

Coupled helixes can be fairly small in radial thickness and the limitations on their use are associated with the t.w.t. helix dimensions at high frequencies and with the relatively long coupling lengths required at low frequencies. The former problem arises because it is necessary to have a diameter ratio of 2:1 or less between the two coupled helixes for efficient broadband use. For typical low-noise tubes at frequences above 2 GHz, the relative dimensions of helix, bulb and support structures, compared with the interaction helix itself, prevent the attainment of such a diameter ratio unless the internal coupling helix is made larger than the interaction helix. At frequencies below about 2 GHz, workers at GEC have shown that it is possible to use an internal coupling helix wound at the same diameter as the interaction helix, but because the pitch must be chosen to enable efficient coupling, this section of the tube is nonactive. The nonactive length increases as the frequency is lowered and the penalty is an increase in size and weight of the final amplifier. In principle, both of the above problems could be solved by using an internal coupling helix wound back over the helix support structure, but the matching and constructional problems would then be comparable to those of direct connection and the performance would be inferior.

The ideal coupling system is that using direct connection, as this offers minimum loss, broadest bandwidth (several octaves if required) and small size. It is widely used by American manufacturers, but it is clear that considerable development work has gone into the design of suitable vacuum seals and impedance transformers incorporated inside the tube. Several British firms are also currently developing direct-connection systems.

2.8 Other characteristics

The above discussion has been on the interrelation of the primary amplifier characteristics, namely noise factor, saturation power, gain, size and weight. For many systems applications it is also necessary to consider other characteristics, such as minimum gain variation with frequency, phase distortion, differential gain and phase between two or more tubes, and the response to environmental conditions such as shock, vibration and temperature.

'Gain flatness' is the only one of these characteristics that accompanies good low-noise design, because of the tendency to use a hollow beam and a large value of γa. One example of this is a tube which operates at 2–8 GHz with a saturation output power of $+3$ dBm, a typical gain of 25 dB and a gain variation of only ± 3 dB over the band. The midband noise factor for this tube is quoted as 5·5 dB. Although the broadband-gain response can be extremely good, it should be pointed out that fine-grain variations with frequency can present a problem in low-noise t.w.t.s. These are caused by helix and attenuator imperfections (particularly significant for low-noise tubes because of the small pitch used) and are usually of the order of ± 1 dB.

The major source of phase distortion for low-level signals is variation of the helix supply voltage. The change in phase is directly proportional to the change in voltage and the total electrical length, and is inversely proportional to the absolute voltage. Since low-noise tubes are usually electrically long and have low helix voltages, the effect is large. For a typical S band tube it would be necessary to stabilise the helix voltage to about $\pm 0 \cdot 10 \%$ to limit phase variations to $\pm 1°$.

Where tubes matched in gain or phase over a given frequency are required, the main difficulties are associated with manufacturing suitably accurate helix assemblies. It has already been mentioned that fine-grain gain variations of the order of ± 1 dB are typical, and as these are associated with random helix imperfections, it follows that, even when the same basic gain response is obtained from two tubes, fine-grain variations as large as ± 2 dB are likely to occur. Very little information is available relating to matched phase variation with frequency, but mismatches due to helix imperfections are again the major problem. Zacharias (1961) has measured differential phase variations within $\pm 2°$ over a 500 MHz band with a C band low-noise tube, but the results included variations as large as 3° in 25 MHz. These results were obtained with a hot match of better than 1·22:1 over the band, and it is probable that a hot match of better than 1·1:1 would be required to limit differential phase variations to less than 1°.

Behaviour under conditions of vibration and shock is determined by relative movements of parts in the tube and of the tube in its focus mount. Because of the small size it is difficult to manufacture a very strong device, and the helix in particular presents severe problems. Alternative slow-wave structures are not suitable for broadband low-noise travelling-wave tubes because of frequency limitations and because of constructional problems associated with the low voltages. It has to be remembered also that a given physical movement results in a larger percentage change in a low-voltage low-noise tube than it would, for example, in a medium-power tube. The solution adopted by many manufacturers is to glaze the helix to its support rods and to use a brazed metal–ceramic gun structure. This sort of gun can have a very small overall diameter, and the smallest possible magnet assemblies are used.

Temperature variations mainly affect the performance of the magnetic focus mount. If steel magnets are used, the effect is not great, but the field strengths in ferrite magnets change by about 0·2 % per degC. This may not be critical if s.f.p.m. mounts with radially magnetised ferrite magnets are used, because although the noise factor varies somewhat as the field strength changes with temperature the focusing should not deteriorate significantly, but it is a major problem with p.p.m. mounts using ferrite magnets. In a p.p.m. stack, the field strength is quite critical and the tube focusing can deteriorate for a fairly small change in the magnetic field. It is therefore necessary to provide temperature compensation for p.p.m. mounts required to operate over a wide temperature range, and this greatly complicates the mount design.

2.9   Intermodulation

In low-noise applications of travelling-wave tubes, with power levels well below saturation, higher-order intermodulation terms than the cubic terms will be negligible. The level of the cubic term will depend upon the exact shape of the transfer characteristic of the tube, and, for a given shape of characteristic, will depend upon the signal levels relative to the saturation level of the particular tube.

To quote a specific measured result (for a tube of a type which might be used in a communication satellite), with two carriers each at 7 dB below saturation level, the power of the $2f_1 - f_2$ intermodulation product is about 24 dB below that of either of the carriers, and of course, this relative level would improve at a rate of 2 dB per dB as the level of the carriers is reduced. The relative level may vary somewhat from tube to tube. The intermodulation behaviour becomes more complex as power levels approach saturation and when more than two carriers are present (Westcott, 1967).

2.10   Tube life

A shortcoming of the low-noise t.w.t. is its fairly short life, usually of the order of a few thousand hours. The main reason for the life being less than that of a medium-power t.w.t. (usually tens of thousands of hours) is the small ratio of active cathode area to the area of other components in the tube. This is a result of the low voltages used and the need for divergent rather than convergent flow in the electron gun, which dictates the use of a small-diameter cathode. In a medium-power t.w.t., on the other hand, it is common to use a beam-area convergence of 20 : 1 or 30 : 1 between the cathode and the helix.

However, improved cleanliness and tube processing coupled with improved understanding and use of cathode activating agents is currently leading to longer tube lives, and just as the lives of medium-power t.w.t.s have been increased from tens to hundreds of thousands of hours for space applications, so the lives of low-noise t.w.t.s are being extended to tens of thousands of hours. One factor which may aid the achievement of long tube life is the provision of preset power-supply units (integral in some cases), as this ensures that the tubes cannot be operated under the wrong conditions. The main argument against the use of preset voltages has been that optimisation of electrode potentials during operation can increase the useful life of the tube. In principle this is true, but in practice it is more likely that the wrong conditions may be set up in the field. This is because independent optimisation of voltages can lead to a noise-factor minimum which is not the lowest minimum. Another danger is that conditions may be optimised when a tube is first switched on, although it often takes some time for the tube performance to stabilise. If the cathode emission varies significantly during the life of the tube, a periodic optimisation procedure may be worth while, but it has now been demonstrated that this is not necessary if the tube has an inherently long life. In fact, one US manufacturer is only prepared to offer attractive life guarantees for tubes with which a preset integral power-supply unit is provided.

The consideration of integral power-supply units cannot therefore be separated from that of tube life. Once a tube with inherently long life has been developed, it follows that the emission will not vary greatly with time, and a preset power-supply unit becomes a valuable safeguard against incorrect operation.

2.11 Possible areas for research and development on t.w.t.s

2.11.1 Fundamental studies

Further experimental investigation of the noise behaviour of t.w.t.s using divergent-flow guns, with particular emphasis on variations in performance with current density and potential profile changes, would be valuable. It would be interesting to apply experience with crossed-field guns to the problem, as this is directly relevant to the operation of a low-noise-t.w.t. gun employing a divergent electric field coupled with a strong axial magnetic field.

Although further theoretical study of the multivelocity nonlinear processes occurring in the cathode region of the gun would be of interest, the problem is so intractable that many assumptions have to be made to obtain results. Because of this, and bearing in mind the theoretical work already done, it is unlikely that practical design information could be derived from such work.

Research into the use of cold cathodes deserves careful consideration, as in principle such cathodes should lead to much improved noise performance. Some work on tunnel cathodes (not directed to low-noise t.w.t.s) suggests that average current densities suitable for lower-power ($\leqslant$ 10 dBm) low-noise tubes are practicable, but very non-uniform emission is currently a feature of these cathodes and it may be necessary to improve this before useful application is possible.

2.11.2 Higher-output-power low-noise t.w.t.s

Development in this area depends largely on being able to use higher current densities from the cathode, and therefore the investigations overlap with some of the work mentioned in the previous section. However, it also depends on the helix voltages and diameters that can be used, and therefore a separate study devoted to extending the power capability of low-noise t.w.t.s might be justified.

2.11.3 Gain and phase variations

A number of systems designers foresee the need for reduced variation in gain and phase response, and for the supply of tubes matched to one another in these characteristics. Some work is already in progress on these aspects, but much remains to be done. The major advance must come through the improvement of helix uniformity and very good broadband matching.

3 OTHER LOW-NOISE ELECTRON-BEAM DEVICES

3.1 Backward-wave amplifiers

In the discussion on t.w.t.s, it was mentioned that any perturbation gives rise to backward as well as forward waves. As its name implies, the backward-wave amplifier makes use of a backward wave for amplification. The interaction mechanism is basically similar to that in a t.w.t. with the exception that the power flow on the slow-wave structure is in the opposite direction to the motion of the beam. The first demonstration of the value of an extended low-velocity drift region in reducing beam noisiness was by Currie (1957) using a backward-wave amplifier. At S band frequencies, noise factors below 4 dB were obtained at a field strength of 1300 Oe, and the rapid development of ultra-low-noise t.w.t.s in recent years has been largely due to application of the principles then outlined.

Despite demonstrations of good noise performance, the b.w.a. has not become widely used for low-noise applications, mainly because of the limited instantaneous bandwidth. Although it is true that the voltage tuning of the operating band gives it advantage over the t.w.t. for certain applications, the gain and noise factor are more variable with frequency than they are in a t.w.t., and the use of the latter device with tunable filters or heterodyne receivers is generally preferred.

3.2 Crossed-field amplifiers

In a crossed-field amplifier a d.c. electric field is applied between a slow-wave structure and the cathode or sole, and a magnetic field at right angles to the electric field constrains the electrons to move in cycloidal paths in the region between. When r.f. fields are applied, the electrons are bunched towards the slow-wave structure, giving up potential energy from the d.c. field to the r.f. wave.

Until recently it was considered that crossed-field devices were inherently noisy, but this view has had to be revised in the light of work reported by Wadhwa and van Duzer (1965). With an S band forward-wave amplifier they obtained a best noise factor of 3·5 dB. The major factors contributing to this result were the development of a short electron gun giving smooth laminar flow, and design to give low diocotron gain (slipping-stream interaction) in the gun and drift regions. The result is particularly impressive in view of the saturation output power of 12 W, and further research into this field should certainly be considered where there is a need for a very wide dynamic range.

3.3 Beam parametric amplifiers

The first successful low-noise parametric beam amplifier was the Adler tube, in which interaction occurs between the fast cyclotron wave and a pump wave. Its low-noise performance depends on the removal of the fast-wave noise by the input coupler, but early tubes were degenerate, using pump frequencies about twice the signal frequency, and this resulted in the introduction of idler noise. A considerable amount of work has been devoted to developing a non-degenerate form of this amplifier, but it has not proved possible to do this without introducing additional beam noise, and the resulting amplifiers are not really better than the latest varactor-diode parametric amplifiers and t.w.t.s. An experimental electron-beam parametric amplifier is shown in Plate 2.

The most interesting form of beam parametric amplifier is probably the d.c.-pumped cyclotron-wave amplifier, in which no r.f. pumping is required as the active coupling between the fast and slow cyclotron waves provides the gain mechanism. It is not possible to extract the slow-wave noise, of course, but the fast- and slow-wave noise temperatures are inversely proportional to the magnetic field strength at the cathode, and therefore a low-noise amplifier is possible in which a very large cathode field is used to reduce noise, with only moderate fields in the interaction region. Adler and Hrbeck (1964) have reported noise factors as low as 1·3 and 2·3 dB at 400 and 1300 MHz, with gains of 20 dB in each case. They suggest that practical devices using cathode magnetic field strengths of 5000 Oe should be possible, with terminal noise factors ranging from about 1 dB at 400 MHz to 6·5 dB at 10 GHz. The performance of these tubes would approach that of the best t.w.t.s, but it is unlikely that they would be more attractive devices.

4 SOLID-STATE MASERS

4.1 History and basic principles

The maser is a microwave amplifier which operates on entirely different principles from those of the tubes described in Chaps. 2 and 3. The word maser is an acronym for 'microwave amplification by stimulated emission of radiation.'† The basic idea of obtaining coherent emission at microwave frequencies by stimulating transitions between two quantum states of a system was suggested by Weber (1953) and was shown to be possible by the classical experiments of Townes and his coworkers (1955) using an ammonia system. Although strikingly successful in demonstrating the validity of the maser principle, the ammonia maser does not provide a practical amplifier, partly because of the low numbers of molecules involved and partly because the frequency of operation is wholly determined by the structure of the molecule and cannot be appreciably influenced by external means.

For these reasons, the possibility of developing a maser based on transitions between the Zeeman levels of paramagnetic ions in solids was attractive. The separation of the Zeeman levels, and hence the operating frequency (given by Einstein's law: $\Delta E = hf$), is determined by a magnetic field acting on the system of ions. Such a solid-state maser is then, in principle, a tunable device, and, because the density of active ions in a solid is greater than in a gaseous system, it is capable of operating at higher power levels than the ammonia device.

Solid-state-maser operation was first demonstrated in a 2-level system—phosphorus-doped silicon—but, in such systems, maser action is necessarily intermittent, and it is difficult to obtain the essential inverted population distribution between the levels.

Bloembergen (1956) proposed the 3-level solid-state maser which is the basis of all the practical solid-state masers which have been

† Perhaps rather ruefully, an alternative basis for the acronym has been suggested, namely, 'means of acquiring support for expensive research!'

realised. Because of the importance of Bloembergen's proposal in the understanding of maser devices, it is worth recalling the essential points. In an assembly of free electrons in a magnetic field, each electron possesses a magnetic moment β, given by

$$\beta = \frac{eh}{4\pi mc}$$

where $\qquad\qquad\qquad e$ = electronic charge

$\qquad\qquad\qquad\qquad\qquad h$ = Planck's constant

$\qquad\qquad\qquad\qquad\qquad m$ = mass of the electron

and $\qquad\qquad\qquad\qquad\; c$ = velocity of light

When acted upon by a magnetic field, the electrons can align themselves with their magnetic moments parallel or antiparallel to the field. The energy difference between these two states is

$$\Delta E = 2\beta H_0 \quad (H_0 = \text{magnitude of the applied field})$$

and, in thermal equilibrium, the relative populations of the two states are determined by Boltzmann's law:

$$\frac{n_2}{n_1} = \exp\left(-\Delta E / kT\right)$$

where $\qquad\qquad\qquad k$ = Boltzmann's constant

and $\qquad\qquad\qquad\qquad T$ = absolute temperature

the lower state (subscript 1) being the more heavily populated. Transitions between the two states are stimulated by application of an electromagnetic field of frequency f such that

$$hf = \Delta E$$

Although upward and downward transitions are equally probable, the actual number of upward transitions which occur with absorption of energy from the electromagnetic field exceeds the number of downward transitions, which are accompanied by emission of energy at the frequency f. There is therefore a net absorption of energy from the electromagnetic field, and this is the basic phenomenon of paramagnetic resonance. The resonant frequency f is linearly related to the applied field and is given by

$$f = \frac{2\beta H_0}{h} = \frac{eH_0}{2\pi mc}$$

On inserting numerical values,

$$f,\ \mathrm{MHz} = 2{\cdot}8H_0,\ \mathrm{Oe}$$

Applied fields in the order of a few thousand oersteds thus correspond to resonant frequencies in the microwave range.

Net absorption occurs only because, in normal circumstances, n_1 is greater than n_2. If, by some means, n_2 is made to exceed n_1, the emission process becomes dominant, and the emitted radiation, being in phase with the stimulating field, will be coherent. There are means of producing population inversion in 2-level systems but, as we have already noted, these are essentially transient. They are discussed in an article by Wittke (1957).

Bloembergen's proposal was to secure a *continuous* inversion of the populations of a pair of levels in a multilevel system, which is characteristic of paramagnetic ions contained in a crystalline host when acted upon by an applied field. The magnetic properties of such ions are rather complicated for both the orbital and spin moments of the unpaired electrons of the ions must be considered, and each ion is acted on, not only by the applied magnetic field, but also by the electric field produced by neighbouring ions in the crystal; but detailed consideration of this complicated situation is not our present object. However, the orbital moment is largely 'quenched' by the crystal field and the resultant magnetic properties of the ions are almost entirely due to the spin (this is of course not true in general, but it is certainly true for the ions in the $3d$ group—the first transition series—which are of present concern). Ions having more than one unpaired spin give rise to more than two energy levels in a magnetic field; four energy states are possible for an ion having three unpaired spins, for example. Just as for two levels, the separation of the energy levels depends on the applied magnetic field, but because of the presence of the crystal electric field the energy-level separation is no longer

linearly dependent on the applied magnetic field (except in certain
special cases) and transitions between each of the various levels are
allowed to various extents. As an illustration, Fig. 7 shows the ground-
state levels of the ion Cr^{3+} in Al_2O_3. Synthetic ruby such as this, having
a ratio of chromium atoms to aluminium atoms close to 0·035 %, is the
most widely used maser material, and in the remainder of this monograph, when the word ruby is used, it is material of this concentration

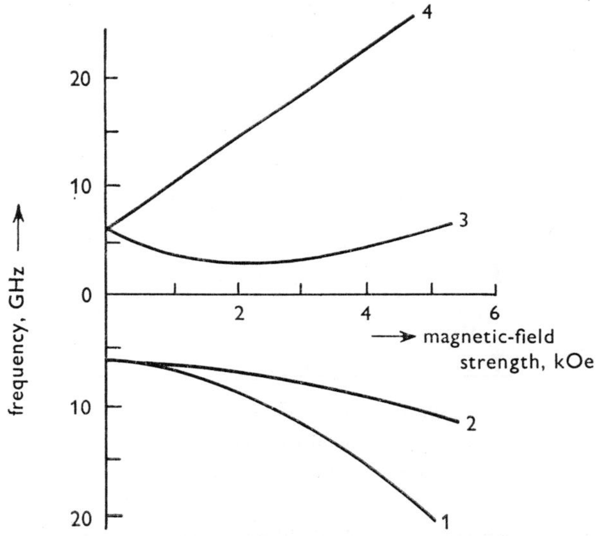

Fig. 7   Energy levels in chromium-doped ruby

which is to be understood. Fig. 7 shows the splitting of the ground-
state levels when the magnetic field is applied at right angles to the
three-fold axis of symmetry of this crystal (*c* axis).

Bloembergen's proposal was to employ a multilevel system such as
this and to obtain a continuous inversion of the populations of a given
pair of levels as a consequence of substantially perturbing the populations of a second pair of levels by the application of electromagnetic
energy at the appropriate frequency.

To see how population inversion can occur in such circumstances
and to establish the significant factors, it is useful to analyse the case
where there are three unequally spaced energy levels as illustrated in

39

Fig. 8. Transitions between these levels occur as a consequence of the interaction between the system of ions and the applied r.f. fields on the one hand and the lattice phonons on the other. The interaction with the lattice causes transitions which result in the creation or annihilation of phonons and are responsible for the maintenance of the Boltzmann population distribution and the spin–lattice relaxation process. These

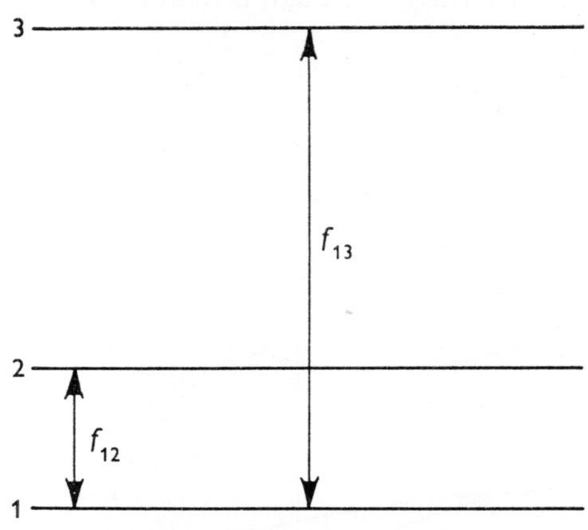

Fig. 8   3-energy-level system
f_{12} = signal frequency
f_{13} = pump frequency

so-called thermal transitions have different probabilities in the upward and downward direction, and if we denote the probability of an upward transition (phonon annihilation) by w_{ij},

$$\frac{w_{ij}}{w_{ji}} = \exp\left(-\Delta E_{ij}/kT\right)$$

Where we have three unequally spaced Zeeman levels and two impressed signals of frequencies f_{12} and f_{13} which stimulate transitions between levels 1 and 2 and 1 and 3 with probabilities W_{12} and W_{13}, respectively ($W_{12} = W_{21}$ and $W_{13} = W_{31}$), the rate of change of the population of level 1 is

$$\frac{dn_1}{dt} = -n_1(W_{13} + w_{13} + W_{12} + w_{12}) + n_2(W_{21} + w_{21}) + n_3(W_{31} + w_{31}) \quad (4.1)$$

A similar equation may be written for level 2, and, since

$$n_1 + n_2 + n_3 = N = \text{a constant}$$

solutions for n_1, n_2 and n_3 can be obtained. When W_{13} is made very large (i.e. the amplitude of the signal at f_{13} is large), the populations of levels 1 and 3 become substantially equal, and in the steady state, when W_{12} is small,

$$\frac{n_2 - n_1}{N} = \frac{w_{23}\{\exp(hf_{23}/kT) - 1\} - w_{12}\{\exp(hf_{12}/kT) - 1\}}{w_{23}\{2 + \exp(hf_{23}/kT)\} + w_{12}\{2\exp(hf_{12}/kT) + 1\}} \quad (4.2)$$

Thus in a 3-level system the population of level 2 can be made to exceed that of level 1 *continuously* on applying a large-amplitude signal at a frequency f_{13} (the 'pump' frequency) if

$$w_{23}\{\exp(hf_{23}/kT) - 1\} > w_{12}\{\exp(hf_{12}/kT) - 1\}$$

or for hf_{23} and $hf_{12} \ll kT$ if

$$w_{23}\, f_{23} > w_{12}\, f_{12} \quad (4.3)$$

In these circumstances, continuous stimulated emission can be obtained from the system at f_{12}, and so amplification of a signal at f_{12} can be obtained. As there are no randomly moving charged particles involved in this amplification process, the device will be, intrinsically, a low-noise amplifier.

The principle of obtaining inversion by this means can be illustrated simply on taking, for example, the system of energy levels shown in Fig. 7, and plotting the populations of these levels against energy for equilibrium at a given temperature, as in Fig. 9. The equilibrium populations of the four levels are substantially different, which implies that the temperature of the system is very low. The application of a signal at frequency f_{13} perturbs, in the first instance, the populations of levels 1 and 3 and, if the amplitude of the signal is made very large, as supposed in this analysis, the probability of stimulated transitions W_{13} greatly exceeds the thermal-transition probabilities w_{13} and the populations of levels 1 and 3 approach equality. In this condition, it is said that the transition is saturated. This does not mean, however, that absorption at f_{13} no longer occurs, but that the power absorbed becomes independent of the signal amplitude. The population distribution following saturation of the 1–3 transition is indicated in Fig. 9, from which it is immediately apparent that, under these conditions, the population of level 2 can exceed that of level 1.

Also represented in Fig. 9 is the saturation of the 1–4 transition. As one would expect, this leads to a substantially larger difference between the populations of levels 2 and 1 than does 1–3 pumping.

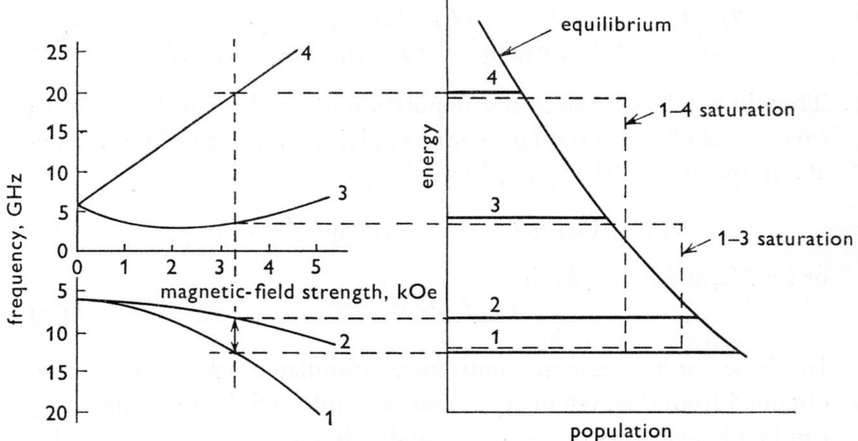

Fig. 9   Population inversion in a 4-level system

Although this illustration gives a useful picture of the inversion process, it is oversimplified by the tacit assumption that the population of level 2 remains unchanged at its equilibrium value despite the application of the pump power. This is unlikely to be strictly true.

A further significant point is that interaction between the ions and the r.f. field occurs over a band of frequencies in the neighbourhood of the resonant frequency. In other words, the paramagnetic resonance lines have a finite width. One reason for this is that, statistically a given ion can only interact with the r.f. field for a finite time before it exchanges its energy with another ion. This is the spin–spin relaxation process and leads to Lorentzian resonance-line shapes. However, there are local random variations in the crystal field and orientation which give rise to what is referred to as inhomogeneous broadening resulting in a Gaussian line shape. It is found in practice that paramagnetic resonance lines generally have shapes somewhere between the two forms.

4.2 Embodiments and design problems

4.2.1. Cavity masers

Bloembergen's proposal was rapidly followed by demonstrations of the feasibility of masers of this type by Scovil, Feher and Seidel (1957) and by McWhorter and Meyer (1958). The latter authors succeeded in measuring the noise temperature of their device and verified the predicted ultra-low-noise performance. Like other early maser devices, these were cavity masers; that is to say the active material was contained in a cavity resonant at both the signal and the pump frequencies. Such devices are inherently regenerative and tend to be unstable, and so they must be employed in conjunction with circulators to separate the input and output. Generally, the introduction of nonreciprocal elements, such as circulators, before the maser results in a degradation of noise performance, which, together with its instability and narrow passband, makes the cavity maser relatively unpractical. Nevertheless, this type of maser has occasionally been employed in operational low-noise systems (e.g. by Jelley and Cooper, 1961).

4.2.2 The travelling-wave maser

Following the early experiments with cavity masers, developments were rapid and received considerable impetus from the discovery of the excellent properties of synthetic ruby as a maser material (Makhov *et al.* 1958) and the construction of a device operating on the travelling-wave principle by De Grasse *et al.* (1959). The travelling-wave maser (t.w.m.) has the following very important advantages over the earlier cavity devices:

(*a*) The t.w.m. is intrinsically not regenerative and therefore does not suffer from the constant gain–bandwidth-product restriction which characterises cavity devices.

(*b*) For a given gain, the instantaneous bandwidth is essentially limited only by the effective width of the paramagnetic-resonance line employed for the signal transition.

(*c*) The device can be made to be completely nonreciprocal by the inclusion of ferrite isolating material.

(*d*) The gain stability is much higher than that of the cavity device.

Almost all operational masers have been of the travelling-wave type and operated at liquid-helium temperatures. Only this type of maser will be considered below. The t.w.m. consists essentially of a microwave structure which can support propagation at both the signal and pump frequencies and which contains the maser material. Practical devices also contain ferrite material to act as an isolator.

If the two Q factors Q_m and Q_0, respectively, are defined in terms of the power P_m emitted by the maser material per unit length and the power P_0 absorbed in the structure per unit length, as follows:

$$\frac{1}{Q_m} = \frac{P_m}{2\pi f_s W_s} \quad \text{and} \quad \frac{1}{Q_0} = \frac{P_0}{2\pi f_s W_s} \tag{4.4}$$

where W_s = stored energy per unit length of the structure, the gain of the t.w.m. can readily be calculated and is

$$G = 27{\cdot}3 \frac{L}{\lambda} \left(\frac{c}{v_g}\right) \left(\frac{1}{Q_m} - \frac{1}{Q_0}\right) \quad \text{decibels} \tag{4.5}$$

where L = length of maser

$\quad\quad \lambda$ = free-space wavelength of signal

and $\quad v_g$ = group velocity of signal on propagating structure.

It can be shown (e.g. Walling and Smith, 1963) that Q_m, the magnetic Q factor, is given by

$$\frac{1}{Q_m} = \Delta n h \gamma^2 g(f_s) \eta M \tag{4.6}$$

where

$\quad\quad \Delta n$ = inverted population difference

$\quad\quad h$ = Planck's constant

$\quad\quad \gamma$ = gyromagnetic ratio

$\quad\quad g(f_s)$ = line shape function

η = filling factor (i.e. the ratio of the r.f. magnetic energy stored in the maser crystal to that stored in the whole structure)

and M = a matrix element determining the transition probability, a number generally between 2 and 4

In practical masers operating at liquid-helium temperatures, values of Q_m of the order of 50 are obtained and so, if reasonable gains are to be obtained from structures of reasonable length ($\sim$ 10 cm), it is apparent from the gain equation, eqn. 4.5, that a slow-wave structure giving a substantial reduction in group velocity ($c/v_g \sim$ 100) must be employed. Suitable structures which allow propagation of the pump frequency, not necessarily in a slow mode, are those consisting essentially of an array of parallel conductors. These structures provide resonant slowing and have the additional advantage that the r.f. magnetic field associated with the slow mode is substantially circularly polarised, permitting the construction of nonreciprocal devices by the inclusion of ferrimagnetic material. The structure which has been most widely employed in t.w.m.s is the comb, but the Karp structure and the meander line have also been employed and may have advantages in some instances. The active material is usually placed alongside the conductors of the structure in such a position that the r.f. magnetic field seen by the material is a maximum. A sectional drawing of a t.w.m., employing a comb structure and designed to operate at a signal frequency of 4·17 GHz, is shown in Fig. 10. Such comb-type slow-wave structures are usually machined from solid copper, although one maser has been described in which the comb structure was deposited by printed-circuit techniques actually on the surface of the ruby (Loriou and Jaouen, 1967).

One of the principal design problems of the t.w.m. is that of calculating the propagation characteristics of slow-wave structures having heavy and nonuniform dielectric loading: the principal information required is the ω/β characteristic and the r.f.-field distribution. Exact analytical solution of this general problem may well be impossible, but approximate techniques have been developed by various authors (e.g. Chen, 1964) which allow some progress in this critical aspect of t.w.m. design to be made. In particular, the occurrence of anomalous modes of propagation (two values of phase constant at a given frequency) can be predicted with good accuracy. A normal ω/β characteristic is shown in Fig. 11 *a*, and an anomalous characteristic in Fig. 11 *b*.

In the anomalous case, at a given frequency two modes of propagation with oppositely directed group velocities occur simultaneously. This is most undesirable as it causes instability and oscillation despite

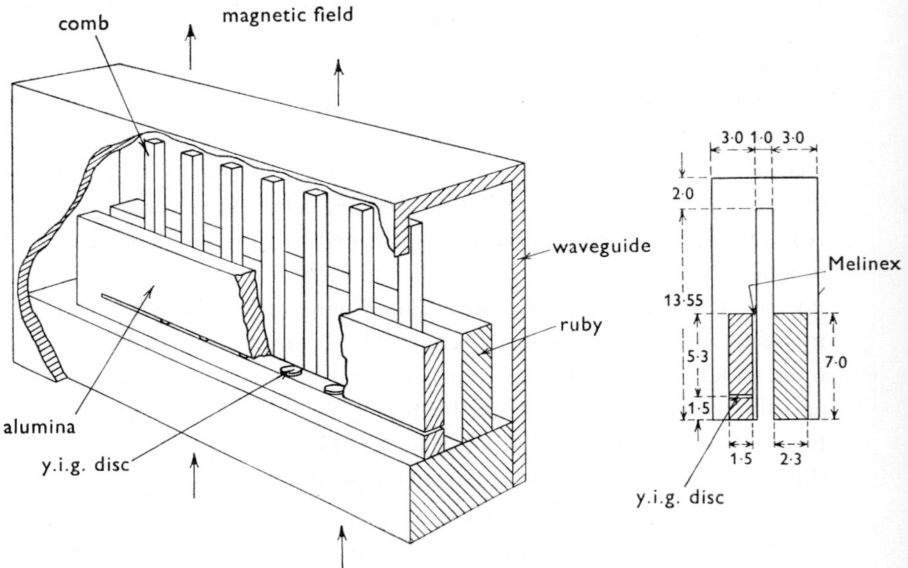

Fig. 10 Diagram of slow-wave structure of a 4·17 GHz travelling-wave maser
Dimensions in millimetres
(Courtesy of *Philips Technical Review*)

the presence of ferrite isolators (De Grasse *et al.*, 1961). Although their occurrence at first caused some bewilderment among maser engineers, anomalous modes are not really mysterious. They tend to occur in structures where the dielectric loading does not extend throughout the space between the conductors and the boundary wall.

The main alternative type of slow-wave structure is the meander line, shown schematically in Fig. 12. Although the slow-wave properties of this structure are basically geometrical, coupling between adjacent sections is also important. The presence of high-permittivity

46

material such as rutile alongside the line will increase the slowing factor. The slowing factor is nevertheless somewhat smaller than can be obtained with a comb structure, but the advantage of the meander

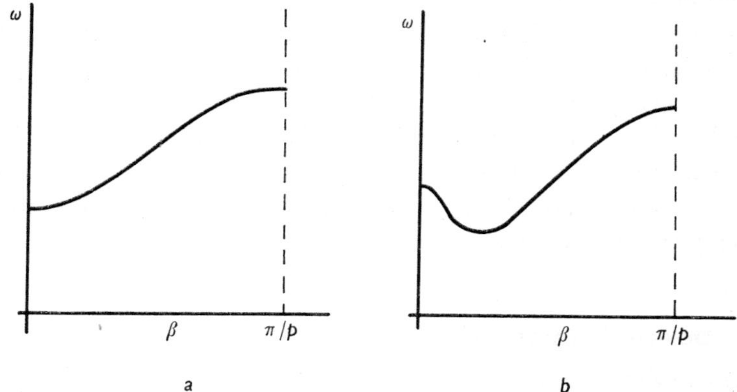

Fig. 11 Propagation characteristics of dielectric-loaded slow-wave structures
a Normal
b Anomalous

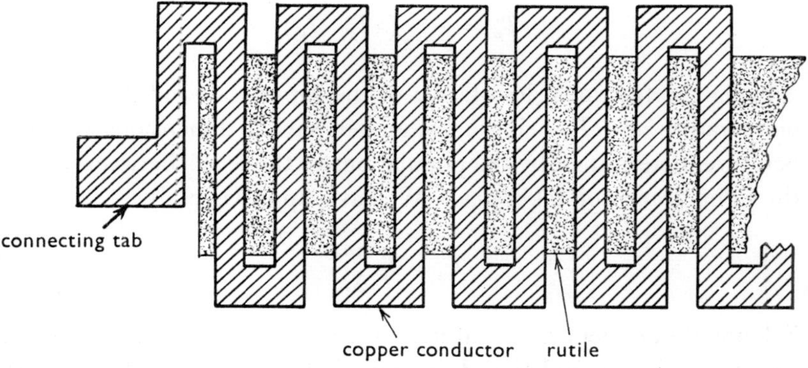

Fig. 12 Schematic of meander line

line is that, unlike the comb, it will propagate signals over a very wide range of frequencies.

The slow-wave structure, then, presents one major t.w.m. design problem, but one which is capable of at least an approximate solution.

The second major t.w.m. design problem is presented by the active material. It is important to know the populations that will result from a particular pumping scheme, and, as already indicated, these are determined by the thermal-transition probabilities between the individual Zeeman levels, and a knowledge of these probabilities is desirable. Theoretical and experimental attacks (Orton *et al.*, 1966, and Stevens, 1967) have shed some light on this exceedingly difficult problem, but it is not yet possible to make a reliable prediction of the population distribution produced by particular pumping schemes in a given material. T.W.M. design is, therefore, in this respect very much an empirical procedure.

4.3   Device characteristics

4.3.1   Noise temperature

The outstanding characteristic of the t.w.m. is its extremely low noise temperature. The equivalent noise temperature of a t.w.m. giving useful gain is, to a very good approximation, given by

$$T_n = \frac{1}{1-\alpha}(\alpha T_\alpha + T_m) \tag{4.7}$$

α being the fractional loss factor of the input lead and T_α its mean temperature and T_m being a parameter having dimensions of temperature, determined by the populations of the levels between which the signal transition takes place. We have, approximately,

$$T_m = T_0 \frac{\Delta n_0}{\Delta n}$$

T_0 being the ambient temperature of the maser crystal and Δn_0 and Δn the equilibrium population differences between the signal-transition levels with and without the pump, respectively. The origins of the two terms in eqn. 4.7 are the loss in the input lead and spontaneous emission in the maser material. Spontaneous emission occurs as a result of random downward transitions from the upper of the signal levels. These are relatively improbable at microwave frequencies and were ignored in the discussion of population inversion (eqn. 4.1), but

48

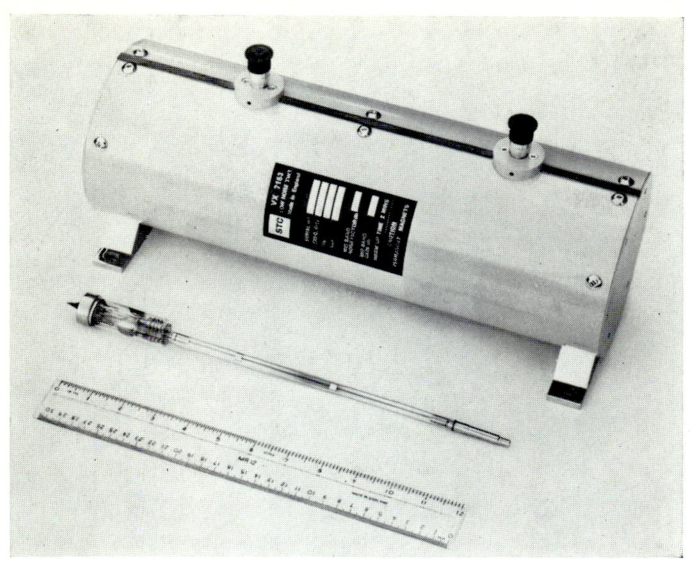

1 S band low-noise travelling-wave tube with radial-magnet
straight-field mount

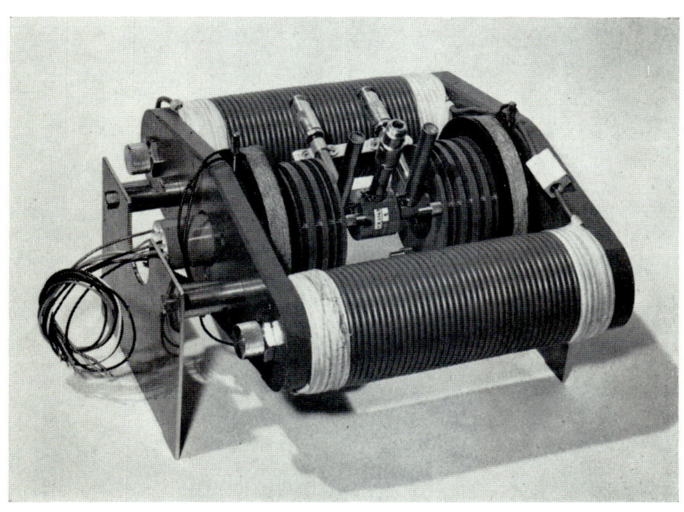

2 Experimental electron-beam parametric amplifier

(Photo: M-O Valve Co. Ltd.)

3 Travelling-wave maser installed at Goonhilly radio station

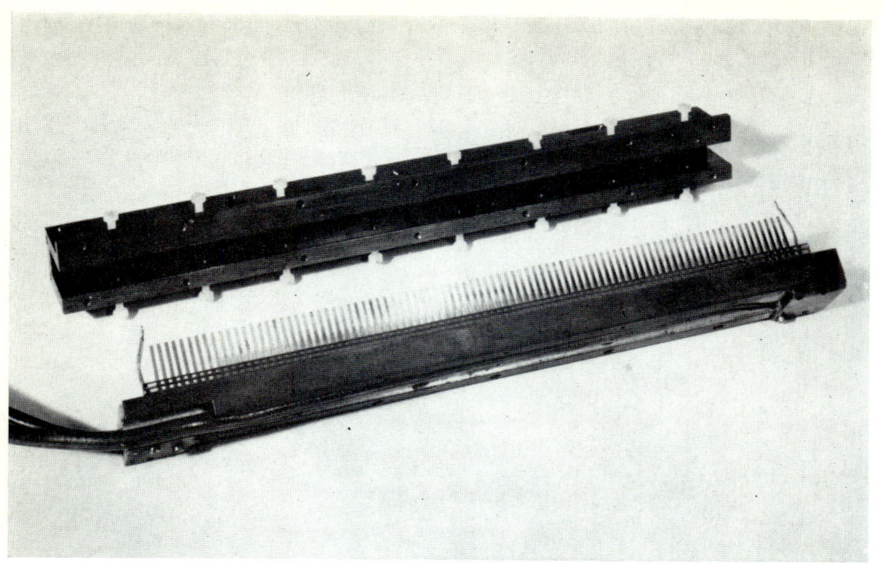

4 Comb structure of travelling-wave maser

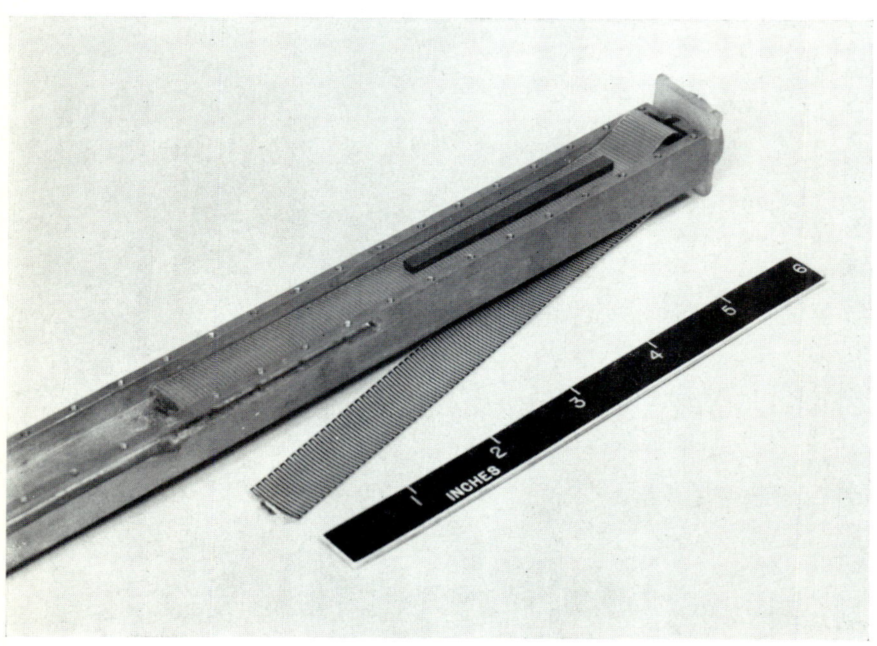

5 Meander-line slow-wave structure

(Photo: L. C. Morris, Radio Corporation of America)

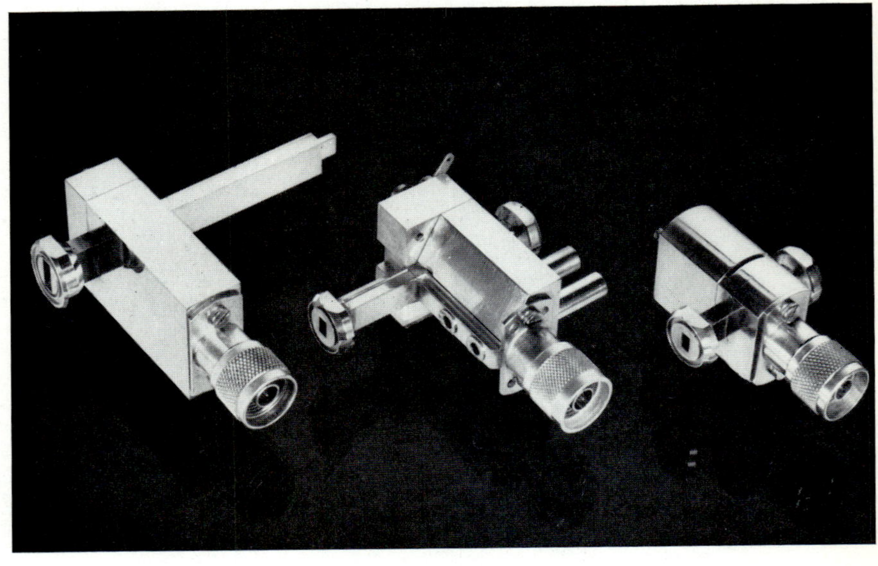

6 Range of parametric amplifiers which use the diode
series resonance to support the idler frequency

(Photo: *Philips Technical Review*)

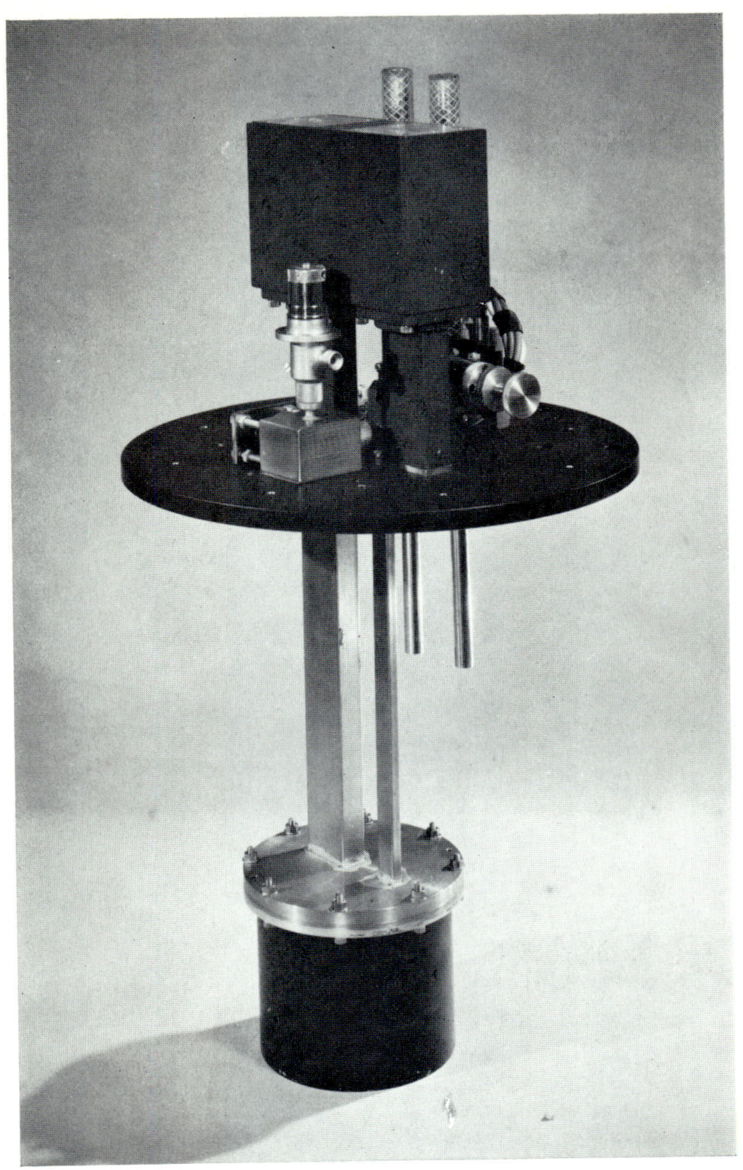

7   Internal structure of liquid-nitrogen-cooled parametric amplifier

(Photo: Ferranti Ltd.)

8   Liquid-nitrogen-cooled parametric amplifier used for
satellite-tracking experiments

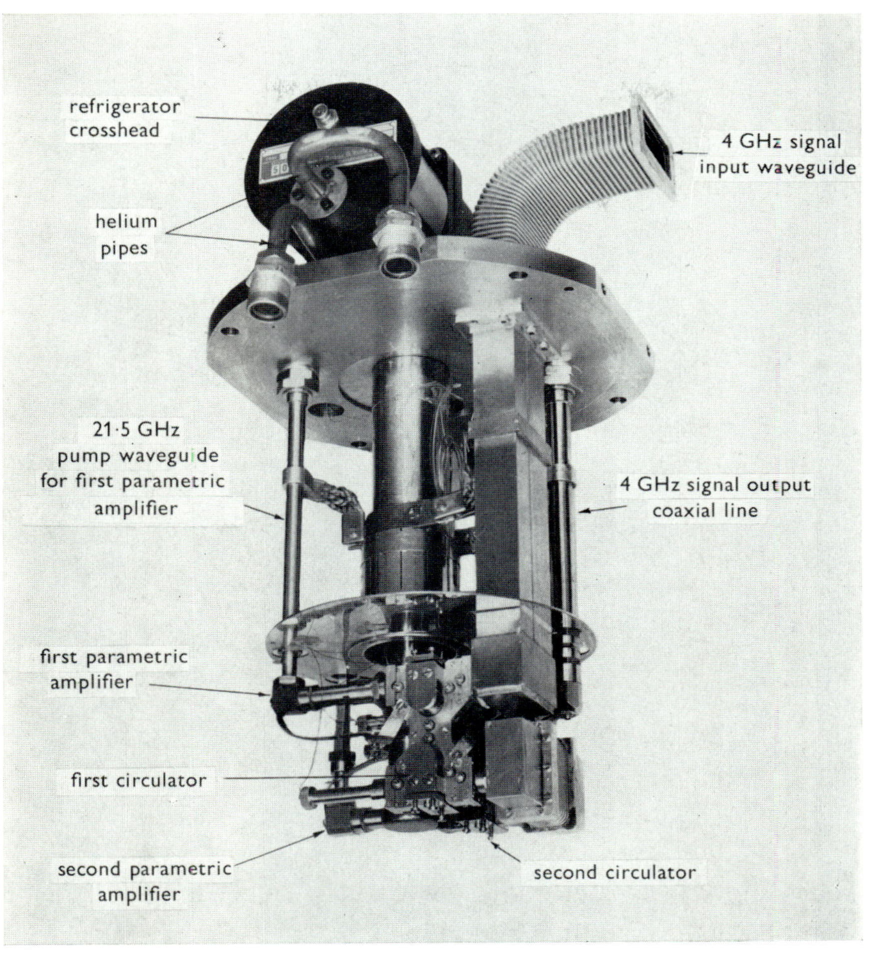

9 Parametric amplifier cooled by miniature closed-cycle refrigerator, as installed at Fucino satellite earth station

(Photo: Prof. G. B. Stracca, G. T. and E., Milan)

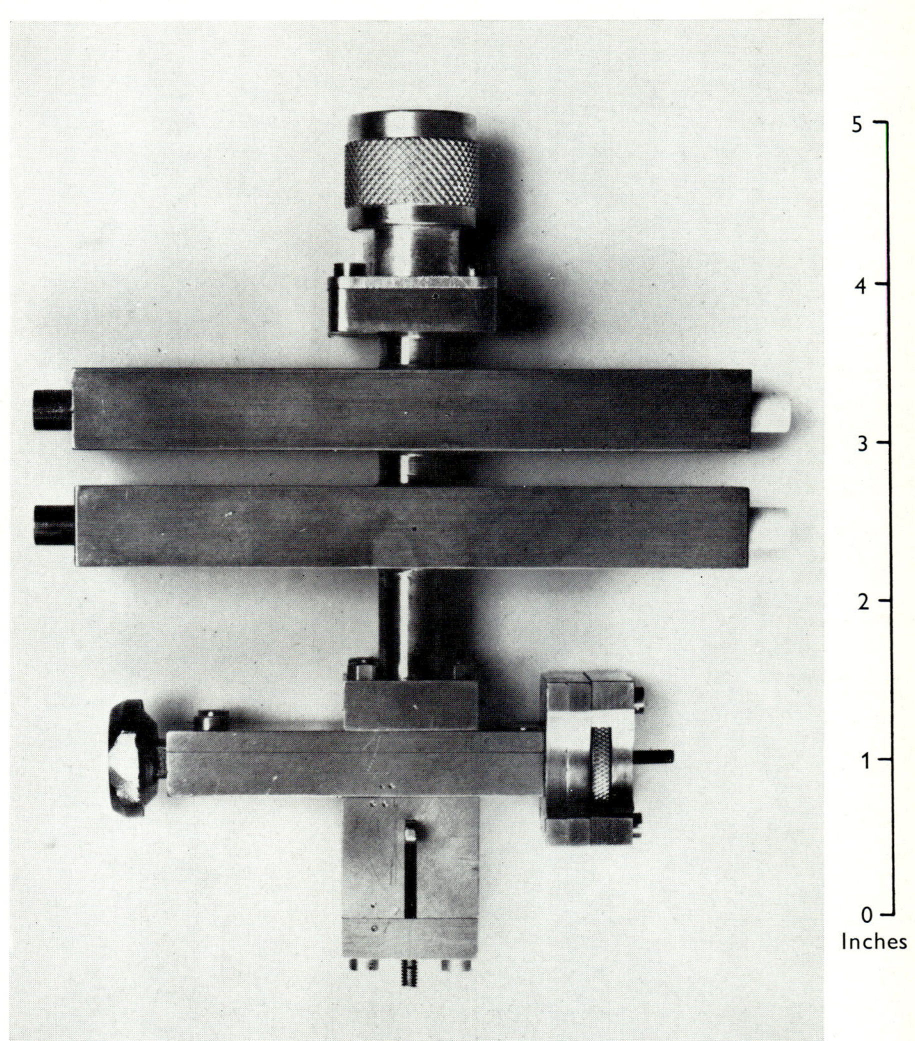

10   Broadband version of simple parametric amplifier, with two
stages of reactance compensation added to signal circuit

are nevertheless significant in noise considerations. With careful attention to the design of the input lead to minimise α and on choosing operating points with high inversion, very low values of T_n may be achieved.

Virtually all the operational devices reported in the literature have measured noise temperatures between 5 and 15 °K, which are in accord with eqn. 4.7.

Eqn. 4.7 is derived for a t.w.m. in which the gain at a given frequency occurs uniformly along the structure. To obtain a broad bandwidth, it may be necessary to stagger-tune the device (Walling and Smith, 1963). This is usually arranged by creating a variation of applied magnetic field along the length of the structure. Field 'staggering' of this kind can result in a degradation of the maser noise performance. A variation of field along the length of the structure is most readily achieved in practice but this degrades the noise performance, and longitudinal sinusoidal variation or transverse variation are to be preferred.

4.3.2　Gain and bandwidth

Factors determining the gain of the t.w.m. have already been discussed, and the bandwidth is the next important consideration.

The frequency dependence of the gain is contained in the line-shape function $g(f_s)$. All paramagnetic resonance lines are broadened by spin–spin interaction processes, by spin–lattice interaction, by crystal inhomogeneities or by all three. The shape is determined by the nature of the process and may be Lorentzian (relaxation broadened), Gaussian (inhomogeneously broadened) or intermediate between the two. The bandwidth of a t.w.m. depends on electronic gain and for a Lorentzian line in a uniform field the relation is

$$B = B_m \sqrt{\left(\frac{3}{G_e - 3}\right)} \tag{4.8}$$

B being the bandwidth to the -3 dB points, B_m the natural width of the resonance line and G_e the electronic gain in decibels. A corresponding expression may readily be derived for the Gaussian line, but for a maser giving a gain in excess of 20 dB, and operating in a uniform field, there is no practical difference between the two. For ruby, the most widely used maser material, B_m is of the order of 60 MHz, and so, for a maser giving an electronic gain of 50 dB, the 3 dB bandwidth is 16 MHz. An instantaneous bandwidth such as this is too low for most practical applications, but it can be increased by variation either of the magnetic field applied to the maser crystal or of the orientation of the crystal with respect to the field.

The type of field staggering employed in practice is dictated by noise considerations and by the gain reduction caused by staggering. Orientation staggering leads to a reduction in the matrix element M (eqn. 4.6) and hence a large sacrifice of gain. Some degree of orientation staggering may well occur owing to imperfections in individual maser crystals, but this is not the normal method chosen to increase the bandwidth of a practical maser.

A single longitudinal step in the field gives the minimum gain reduction but increases the noise. It is, however, readily achieved in practice either by shimming or by making use of one or more auxiliary coils, and has been widely used. In ruby masers, bandwidths of up to 60 MHz are realisable without the need to use more than one pump frequency.

In Fig. 13, the variations of gain and bandwidth for a maser having an electronic gain of 50 dB and a bandwidth of 16 MHz in a uniform field, are plotted for three different types of field staggering.

4.3.3 Power handling and dynamic range

The lower limit of the dynamic range of a t.w.m. is determined by the internal noise level. For a t.w.m. with a bandwidth of 50 MHz and an equivalent noise temperature of $10\,°K$, the lowest detectable input signal is 10^{-15} W (-120 dBm). On the high side, the limit is set by the gain saturation which occurs as the population difference between the signal transition levels begins to be reduced by the stimulated transitions. For a t.w.m. with negligible structure loss and giving a small-signal net gain of G_0 dB, the gain G in decibels at an output power level P_0 is given to a good approximation by

$$\frac{G_0 - G}{G_0} = \frac{P_0}{P_s L} \tag{4.9}$$

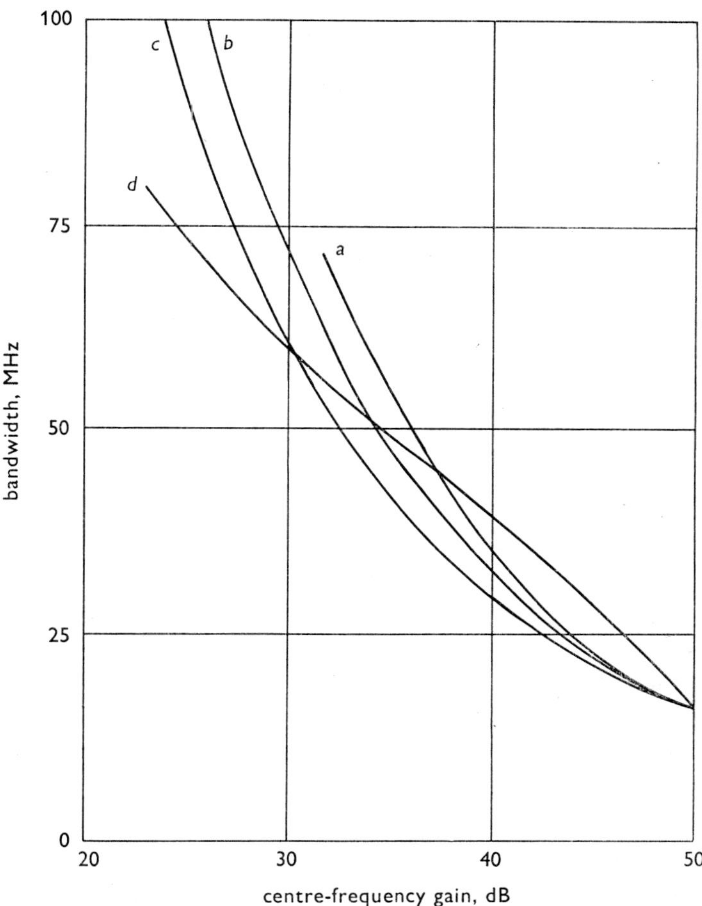

Fig. 13   3 dB bandwidth as a function of gain at centre frequency with various broadbanding techniques (unmodified gain = 50 dB)

  a   Single-step variation of magnetic field
  b   Sinusoidal variation of magnetic field
  c   Linear variation of magnetic field
  d   Unmodified maser followed by gain-equalisation circuit

(Courtesy of *Philips Technical Review*)

Here P_s is the maximum power available per unit length of the maser
and is the product of the zero-signal inverted population-density
difference Δn, the volume of crystal per unit length and the signal
quantum divided by the effective spin–lattice relaxation time τ_1, i.e.

$$P_s = \frac{\Delta n h f_s V_c}{\tau_1} \qquad (4.10)$$

So, for a maser giving a small-signal gain of 30 dB, the gain is reduced
to 27 dB when the output power is $P_s L/10$. For an S band maser
employing ruby as the active material

$$\Delta n = 7 \times 10^{17}\,\text{cm}^{-3}$$

$$V_c L = 2 \cdot 4\,\text{cm}^3$$

and
$$\tau_1 = 100\,\text{ms}$$

and so $P_0 \simeq 2 \cdot 8 \times 10^{-6}\,\text{W} = -26$ dBm. This figure is substantiated by
experiment (Tabor, 1964), and thus the dynamic range of such a t.w.m.
is from -120 dBm to -53 dBm.

Masers operating at higher frequencies saturate at higher power
levels because of the larger Δn and signal quantum $h f_s$ and shorter τ_1,
although these effects are offset to some extent by the fact that the
volume of active crystal is small, but nevertheless a Q band t.w.m.
(Arams and Peyton, 1965) exhibits a 3 dB gain compression at an output power of -15 dBm.

It is important to realise that the saturation behaviour so far
discussed relates to situations in which the input power is changed
slowly, i.e. at a rate substantially slower than the spin–lattice interaction
rates, so that the populations continuously readjust themselves to
quasiequilibrium values under the combined effects of the signal, pump
and spin–lattice relaxation processes. Under pulsed conditions, gain
saturation occurs at peak-power levels which exceed the c.w. values by
the duty cycle (it being understood that the pulse length is substantially less than the spin–lattice interaction time).

Tabor (1964) has published the results of measurements on the c.w.
and pulsed gain-saturation characteristics of a t.w.m. These are
summarised in Fig. 14.

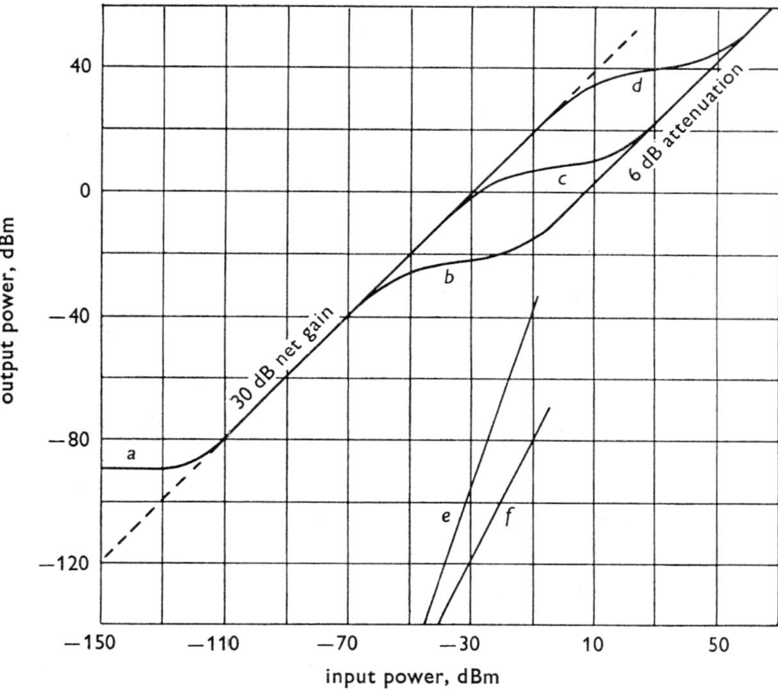

Fig. 14   Output against input for travelling-wave maser illustrating noise level, gain saturation dynamic range and intermodulation distortion for both c.w. and pulse operation

    a   Maser amplifier, $T_n = 3{\cdot}5\,^{\circ}\text{K}$
    b   C.W. operation
    c   30 dB pulse-duty ratio (radar)
    d   63 dB maximum pulse-duty ratio
    e   Intermodulation
    f   Generation of second harmonic

(Courtesy of the Institute of Electrical & Electronics Engineers)

4.3.4 Linearity and intermodulation

Because of the effects discussed in the preceding section (i.e. the difference in behaviour with slow and fast changes in power level), the t.w.m. behaves as a linear amplifier even when operating with compressed gain. This is a remarkable and valuable feature.

Intermodulation distortion occurs, not through a gain-saturation mechanism as in other amplifiers but because of spin–spin coupling in the maser material in a manner analogous to harmonic generation and frequency mixing occurring in ferrite devices. The magnitudes of the effects are determined by the spin–spin relaxation time τ_2 (about 5×10^{-9} s for ruby). The effects have been calculated by Schulz du Bois (1964) for a 4·17 GHz ruby t.w.m., and he finds that the second-harmonic power generated (in dBm) is given by

$$P(2f_1) = 2P(f_1) - 103\cdot5 \qquad (4.11)$$

and the intermodulation power (also in dBm) by

$$P(2f_1 - f_2) = 2P(f_1) + P(f_2) - 92 \qquad (4.12)$$

These equations assume a linearly polarised r.f. magnetic field at the signal frequency and thus probably represent an upper limit, since in practice there is an elliptically polarised transition driven by an elliptically polarised r.f. field.

Under c.w. conditions, the output power of a maser is usually about -30 dBm. Thus the second harmonic generated will be at a level of $-163\cdot5$ dBm and the intermodulation product (assuming the power levels at f_1 and f_2 to be of similar magnitude) will be at -182 dBm. Both these magnitudes are far below the detectable-signal limit of the maser (about -90 dBm output).

Under pulsed conditions, intermodulation and harmonic generation may be observed, as indicated in Fig. 14, but it is found that the magnitudes of the effects are somewhat less than calculated, supporting the suggestion that the expressions quoted represent upper limits.

4.3.5 Recovery after overload

The effect of a very-large-amplitude signal is to saturate the maser to a degree dependent on the amplitude and duration of the pulse signal. In the saturated condition, the maser material is completely trans-parent and therefore the device cannot be damaged by the application of large-amplitude signals. On removal of the signal, the quasi-equilibrium population distribution (and hence the gain) is restored in a time determined by the spin–lattice interaction rates, which, for a ruby maser, will be a few tenths of a second. For many purposes this

54

is intolerably long and this is perhaps the most undesirable characteristic of the t.w.m. It is relatively unimportant in communication applications but very significant in any possible radar applications.

4.3.6 Response to spurious signals

Signals at frequencies outside the limits of the passband of the maser slow-wave structure are not propagated. Furthermore, signals lying within the structure passband but outside the width of the paramagnetic resonance line do not affect the maser operation. For all practical purposes, the maser is immune to signals at frequencies lying beyond the instantaneous response characteristic.

4.4 Practical maser devices

Cavity masers have been used for radio astronomy and deep-space radar experiments, but the majority of operational devices are of the travelling-wave type. Early versions used simple comb structures, large external permanent magnets and liquid-helium cooling. The substitution of small persistent-current superconducting solenoids for the external magnet greatly increased the amplifier stability and reduced the weight. An example of an operational maser is shown in Plate 3. This maser was developed for satellite-communication use by the Mullard Research Laboratories and is installed in the British Post Office Earth Station at Goonhilly Downs, Cornwall. It employs a superconducting magnet which is of relatively light weight and has a highly stable and uniform field (Hentley, 1964). The active material is synthetic ruby. The slow-wave comb structure is shown in Plate 4. It is basically similar to that shown in Fig. 10, although the inert alumina dielectric loading has been replaced by additional ruby. Yttrium–iron-garnet discs provide adequate attentuation in the backward direction.

The magnetic field of 3250 Oe is applied at $90°$ to the c axis and the pump transition is between levels 1 and 4 (Fig. 9) at a frequency of 30·15 GHz, the signal transition being between levels 1 and 2 at 4·17 GHz. To increase the bandwidth of the maser, a 'step' of about

10 Oe is introduced in the magnetic field, and the performance of the device is then as follows:

gain	32 dB
bandwidth	55 MHz
forward loss	$\sim$ 30 dB
reverse loss	$\sim$ 65 dB
noise temperature	8 °K
operating temperature	$1\frac{1}{2}$ °K

Sophisticated masers having essentially similar characteristics have been described by workers at the Bell Telephone Laboratories (DeGrasse *et al.*, 1961, and Tabor and Sibilia, 1963) and the Airborne Instrument Laboratories (Okwit and Smith, 1962).

Recently, attention has been given to the construction of masers with large instantaneous bandwidths, and, on using elaborate field-staggering techniques, bandwidths in excess of 200 MHz with acceptable gains have been realised. As an example, a 5 GHz maser has been described (Du Gruyl, Okwit and Smith, 1965, and Okwit, 1967) which has an instantaneous bandwidth of 260 MHz and a net gain of 31 dB. This performance is only achieved by using 16 in of active length, four pump frequencies, operation at 2·1 °K a and magnetic-field gradient in the axis parallel to the teeth of the comb structure. This type of field variation has the advantage of producing the minimum increase in overall noise temperature.

The principle of stagger tuning can be extended so that the instantaneous bandwidth of the maser is comparable with the total-structure passband. This would, however, lead to a cumbersome device, possibly with rather undesirable noise characteristics. For most maser applications it is preferable to limit the instantaneous bandwidth and to provide a facility for tuning.

The centre frequency can be tuned within the passband of the maser structure by altering the applied magnetic field together with the pump frequency. Tunable travelling-wave masers have been built with 'single-knob' tuning, and obviously if the pump is a b.w.o. this presents no great difficulty. It is also possible to find operating points for which the pump frequency is nearly constant over a wide range of signal frequencies and fields, and these may be exploited to advantage in tunable applications.

The tunable bandwidth of a t.w.m. cannot be increased indefinitely, but is limited by the passband of the slow-wave structure, and the structure alterations needed to increase the bandwidth inevitably increase the group velocity and hence decrease the gain. The meander line is more suitable in this situation than the comb, and wide tuning ranges have been reported by Morris using a meander line and Cr^{3+} in TiO_2 as the active material. The flexible meander line shown in Plate 5 is made by printed-circuit techniques using Mylar as substrate (Morris and Miller, 1965). The following performance has been described:

tuning range	2–3 GHz
gain	23 dB
noise temperature	8 °K
pump frequency	49–52 GHz

To some extent, instantaneous bandwidth and tuning range can be interchanged by adjusting the magnetic-field staggering. For example, a similar maser operating in the 4–4·2 GHz band can be adjusted to have a 50 MHz passband tunable over a 500 MHz range or a 180 MHz passband tunable over a 100 MHz range. The noise temperature rises from about 7 to 10 °K as the bandwidth is increased.

An ingenious alternative solution to the problem of obtaining tunable maser performance has been reported by Okwit (1967). In this scheme a 6 GHz t.w.m. is preceded by a cooled-diode parametric up-convertor, which can be nearly noiseless, and followed by a down-convertor (Fig. 15). Both parametric devices have a common pump which is tunable and thus both input and output are at the same frequency. Okwit reports a tuning range of 1·6–2·4 GHz with an overall noise temperature of 15 °K.

4.5 Environment, life and ancillary requirements

All present-day operational masers operate at liquid-helium temperatures, and this low-temperature environment is provided either by liquid helium contained in a large Dewar vessel or by a continuously operating helium refrigerator as indicated in Fig. 16. The latter has a number of advantages in practice.

A magnetic field is required, and, if this is supplied by an external permanent magnet, temperature control of the magnet is necessary. The majority of masers, however, now employ persistent-current superconducting magnets. As these are screened and in the low-temperature environment, external temperature control is unnecessary.

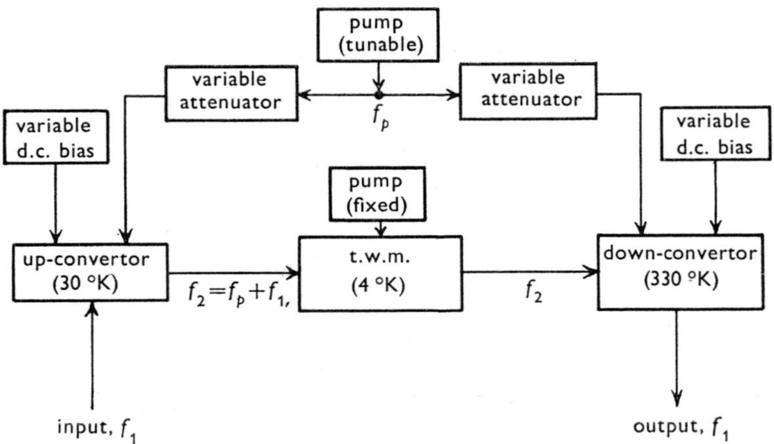

Fig. 15   Up-convertor–maser–down-convertor combination
(Courtesy of S. Okwit, Airborne Instruments Laboratory)

It is necessary to supply pump power to the maser, and the pump frequency and the power level will depend on the maser crystal and the operating point employed. Although coherent monochromatic pumps are generally used, neither of these conditions is essential, and indeed maser operation has recently been reported in Tm^{2+} in CaF_2 (Sabisky and Anderson, 1966) when selective depopulation of one of the ground-state levels has been secured by pumping with incoherent poly-chromatic polarised *optical* radiation.

The life of the maser is virtually unlimited, such failures as have been reported being mechanical failures due to thermal cycling, and these are usually repairable.

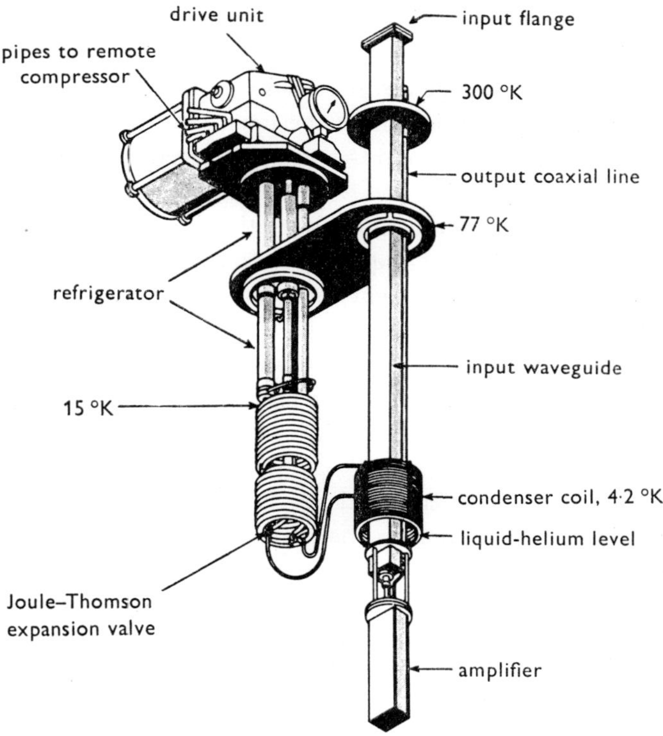

Fig. 16   Maser mounted on a closed-cycle refrigerator, as used at Andover Earth Station

(Courtesy of Bell Telephone Laboratories)

4.6 Summary

The solid-state travelling-wave maser has shown itself to be a viable practical device with the following unique attributes:

(*a*) The noise performance is close to the ultimate attainable.

(*b*) It has a high degree of linearity even under gain-compression conditions, and its intermodulation and crosstalk characteristics are unique.

(*c*) It has 'inbuilt' nonreciprocity and its gain stability is very good.

(*d*) It is, compared with the parametric amplifier, insensitive to variations in pump frequency and amplitude.

(*e*) It is unaffected by signals at frequencies outside its response characteristics.

(*f*) It cannot be damaged by overload signals.

These advantages, which make the device outstandingly good in communication systems, are offset to some extent by the following disadvantages:

(i) Liquid-helium-temperature operation is necessary.

(ii) A magnetic field is required, although, since there must be a liquid-helium-temperature environment, persistent-current super-conducting magnets may be employed with advantage.

(iii) Gain saturation is apparent at low input-power levels.

(iv) Recovery from overload is slow.

(v) Large instantaneous bandwidths and tunability are not readily achieved.

(vi) The device is not easy to design and construct and it also puts rather stringent conditions on the perfection and uniformity of the single paramagnetic crystals employed.

There are few applications in which the outstanding noise characteristics of the maser can be fully utilised, and, in general, systems engineers are not willing to consider the use of liquid-helium-cooled devices unless the ultimate in noise performance is required. Attempts so far made to assess the possibilities of constructing a maser with a satisfactory performance at temperatures higher than those of liquid helium have been discouraging, and the possibilities of optical pumping, which it was once thought might make room-temperature maser operation feasible, do not seem very bright. It seems, therefore, that the maser is irredeemably a liquid-helium-temperature device.

5 VARACTOR-DIODE PARAMETRIC AMPLIFIERS

5.1 Introduction

A parametric amplifier is a device based on the properties of a non-linear reactance—usually capacitative. In a linear capacitance, the charge stored is proportional to the applied voltage, but it is also possible to have a capacitance in which the charge is not linearly related to the applied voltage. Then, when several alternating voltages are applied simultaneously to the capacitance, frequency mixing takes place and the capacitance is capable of transferring energy from one frequency to another. In the parametric amplifier, this property is used to take energy from a 'pump' source at a high frequency to amplify a signal at a lower frequency.

The first practical realisation of this principle was actually based on a nonlinear inductance rather than a capacitance, with the non-linearity depending upon the properties of ferrite material (Weiss, 1957). Unfortunately, large levels of pump power were required to provide the necessary nonlinearity. With pump levels in the kilowatt region, this type of amplifier could have only very limited applications. The amplifiers to be discussed here are all based on the use of a non-linear capacitance in the form of a semiconductor p–n junction. The design and performance of the amplifiers themselves will be considered first and the details of the diode itself will be described in the concluding section.

When a semiconductor junction is used as an electronically variable capacitance in a parametric amplifier, it is essential that virtually no current should flow across the junction. Any current which does flow inevitably generates 'shot' noise, and such a current would therefore degrade the amplifier performance. In the absence of such shot noise, an ideal device, depending purely upon capacitance for its amplifying action, would not generate any noise. However, a real diode is not purely a capacitance as the resistivity of the semiconducting material away from the neighbourhood of the junction is not negligible, and so there is always a series resistance associated with the capacitance. Thus, even if shot noise is eliminated, resistance or Johnson noise must inevitably arise in this series resistance within the diode. The effective noise level will depend upon both the signal and pump frequencies as well as upon the properties of the diode, and, as it is thermal in orgin, the noise level also depends upon the temperature of the diode.

The theoretical variation of the capacitance of a p–n junction, upon which the operation of the parametric amplifier depends, can be described approximately by

$$C_j(V) = \frac{C_0}{\left(1 - \dfrac{V}{\phi}\right)^{1/n}} \tag{5.1}$$

where $\qquad C_0$ = capacitance at zero bias

$\qquad\qquad\quad V$ = bias potential

$\qquad\qquad\quad \phi$ = contact potential

and where $\qquad n = 2$ for an abrupt junction

or $\qquad\qquad n = 3$ for a linearly graded diffused junction

As the capacitance of the diode varies with the applied voltage, the maximum capacitance C_{max} is determined by the onset of significant forward current. The limit is usually quoted as $I_{for} = 1$ μA, although, for cooled amplifiers, 0.1 μA would be more appropriate. The minimum capacitance C_{min} is usually quoted at a reverse bias of 1 V.

In an amplifier, the capacitance is modulated by the application of microwave power at the pump frequency, and

$$\frac{1}{C_j} = \frac{1}{C_0}(1 + 2\gamma \cos 2\pi f_3 t) \tag{5.2}$$

where f_3 is the pump frequency, and higher terms are usually ignored, C_0 is the capacitance without the pump power, and γ is the diode-capacitance modulation coefficient. γ is, of course, a function of the voltage developed across the diode at pump frequency, but will have a maximum attainable value defined by

$$\gamma = \frac{C_{max} - C_{min}}{2(C_{max} + C_{min})} \tag{5.3}$$

where C_{max} and C_{min} are as defined above. When γ is referred to, it is usually this maximum value which is intended. There are other versions of the definition of γ, but in most circumstances the various definitions are at least approximately equivalent to each other.

If only the pump and signal frequencies are present, there cannot be a steady flow of energy from one to the other, except when the pump frequency is exactly twice the signal frequency. To attain the required steady flow of energy, it is necessary for the resultant voltage across the capacitance to be maintained in the correct phase, and this can be achieved only by permitting power to flow at a third frequency. This is the 'idler' frequency which is equal to the difference between the signal and pump frequencies. The amplifier therefore consists of a series of frequency-selective circuits which will permit power to flow only at these three frequencies, and the nonlinear capacitance forms the common element between the tuned circuits. More complex forms of amplifier can be designed, however, in which additional frequencies are also present. The theoretical power flow into and out of an idealised nonlinear capacitance is described in terms of two generalised equations known as the Manley–Rowe relations, which can be written (Manley and Rowe, 1956) as

$$\left. \begin{array}{l} \displaystyle\sum_{m=0}^{\infty} \sum_{n=-\infty}^{\infty} \frac{m\,P_{mn}}{mf_1 + nf_2} = 0 \\[1.5em] \displaystyle\sum_{n=0}^{\infty} \sum_{m=-\infty}^{\infty} \frac{n\,P_{mn}}{mf_1 + nf_2} = 0 \end{array} \right\} \tag{5.4}$$

P_{mn} represents, algebraically, the power flow into the nonlinear capacitance, and f_1 and f_2 are the frequencies at which power is fed to the capacitance.

If we consider 3-frequency amplifiers, the three frequencies being

f_1, f_2 and f_3, where $f_3 = f_1 + f_2$, the general Manley–Rowe equations can then be simplified to

$$\frac{P_1}{f_1} + \frac{P_3}{f_3} = \frac{P_2}{f_2} + \frac{P_3}{f_3} = 0 \tag{5.5}$$

Provided that internal losses are ignored, this equation can be used to understand the operation of some of the different types of parametric amplifier.

There are three basic modes of operation which are important:

(a) The 'sum-frequency' or 'upper-sideband-convertor' mode

In this mode, f_1 is the signal frequency and power is supplied from an external source at frequency f_2. The output frequency is f_3, as

$$\frac{P_1}{f_1} = -\frac{P_3}{f_3}$$

and if power into the reactance is defined as positive, P_3 must represent power leaving the reactance. The maximum power gain, ignoring circuit losses, is

$$\left| \frac{P_3}{P_1} \right| = \frac{f_3}{f_1} \tag{5.6}$$

The limited gain of such an amplifier restricts its applicability to relatively low frequencies if the pump frequency is not to be inconveniently high.

(b) The 'lower-sideband-convertor' mode

In this mode, f_1 is the signal frequency, but power is supplied at frequency f_3. Hence, P_3 is positive and both P_1 and P_2 must be negative; so the circuit can deliver energy at frequencies f_1 and f_2. In the lower-sideband-convertor mode, power is extracted at f_2. (If the frequencies of the lower-sideband-convertor mode are so chosen that f_2 is greater than f_1, the device is sometimes called a difference-frequency up-convertor.)

(c) The 'difference-frequency reflection amplifier'

This mode is basically the same as (b) except that the output is at the signal frequency f_1, and the amplifier is a single-port device. In the latter two modes, the behaviour of the amplifier at frequency f_1 can be described in terms of a 'negative resistance' $-R$ which will appear in the equivalent circuit of the amplifier at this frequency. In the difference-frequency reflection mode, the negative resistance appearing in the signal circuit at resonance is given by the equation (Blackwell and Kotzebue, 1961)

$$-R = -\frac{\gamma^2}{\omega_1\,\omega_2\,C_j^2 R_{T2}} \tag{5.7}$$

where R_{T2} is the total series resistance at frequency f_2.

For some applications, a combination of the up-convertor and negative-conductance modes has been used, with part of the signal being taken at f_1 and part at f_2.

The lower-sideband-up-convertor mode may be used as a single amplifier or combined with a down-convertor (usually of the mixer type), with the pump frequency of the up-convertor and the local-oscillator frequency of the down-converter derived from a common oscillator (with an offset introduced, equal to the required output frequency).

In general, however, the 1-port reflection amplifier is used for most practical low-noise applications. As the input and output signals are at the same frequency, an additional circuit element is necessary to separate input from output. In these amplifiers, f_3 is usually referred to as the 'pump' frequency, f_2 as the 'idler' frequency and f_1 as the 'signal' frequency.

Amplifiers using either of the negative-resistance modes of operation, (b) and (c) above, are very sensitive to changes in circuit impedance and to fluctuations in the power level and frequency of the pump source. Nonreciprocal devices such as isolators and/or circulators are used to protect the amplifier against changes in impedance at the aerial or at the following stage of the receiver. For stability, the gain is usually limited to 20–25 dB for a single-stage device. If a higher gain is required, a number of 10–15 dB stages are connected in cascade.

A wide range of amplifier designs are available and the choice will depend on the particular application. In many applications, the lower limit of background noise in the system is determined by thermal noise entering the aerial from the terrestrial surroundings, and for such applications, provision of a very-low-noise-temperature receiver would be uneconomical, and relatively simple parametric amplifiers may be entirely acceptable. For the reception of the weak signals from communication satellites or interplanetary probes, however, extremely low system noise temperatures are essential, and specialised parametric-amplifier designs have been developed having noise temperatures comparable with those of the masers described in Chap. 4.

Bandwidth requirements also vary. For some terrestrial communication and radar requirements, bandwidths of a few megahertz are adequate. Although early satellite-communication systems were restricted to a few tens of megahertz, the growth of international traffic will make it necessary to be able to receive signals over the whole 500 MHz band which has been allocated for civil satellite systems. For military applications it is possible to envisage systems requiring parametric amplifiers with even larger bandwidths, although alternatively it may be necessary to provide a relatively narrow instantaneous bandwidth electronically tunable over an extremely wide frequency band.

Methods of meeting these varied requirements are discussed in the following sections.

5.2 Diode equivalent circuit and amplifier noise temperature

In order to calculate the noise temperature of the parametric amplifier, it is necessary to use an equivalent circuit which will take into account the series or 'spreading' resistance of the diode, which is a source of Johnson noise at both signal and idler frequencies. The diode equivalent circuit is shown in Fig. 17a. It consists of the junction capacitance C_j in series with resistance R_s, and in parallel with this combination is an internal stray capacitance C_{p2}. In series with this combination is the lead inductance L_p with the capacitance due to the encapsulation in parallel with this series circuit. However, for most purposes, it is acceptable to combine C_{p2} with the other elements to give the simplified equivalent circuit of Fig. 17b. It can be seen that this equivalent circuit has both a series and a parallel resonance. These

66

resonances are very important in the design of high-performance amplifiers.

It is convenient to introduce the spreading resistance R_s into the calculations in the form of a cutoff frequency, defined as

$$f_c = \frac{1}{2\pi R_s C_j} \tag{5.8}$$

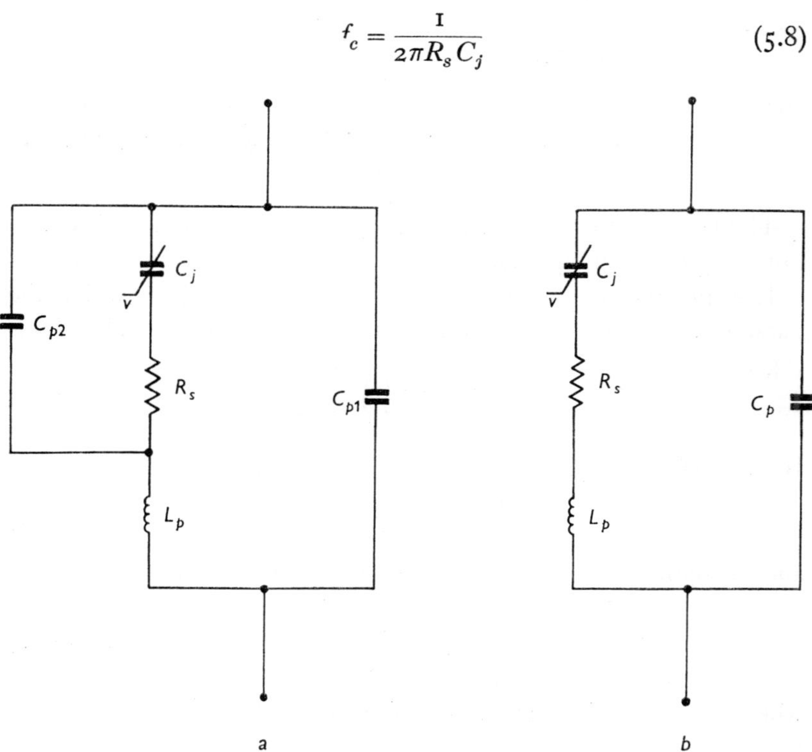

Fig. 17   Equivalent circuits of varactor diodes

The noise temperature for a negative-resistance reflection-type parametric amplifier can be calculated from the following equation (Uenohara and Seidel, 1961):

$$T_n = T\left(1 - \frac{1}{G}\right) \frac{\left(\dfrac{f_c}{f_2}\right)^2 \dfrac{\gamma^2}{1+\alpha} + 1}{\dfrac{f_c^2}{f_1 f_2} \dfrac{\gamma^2}{1+\alpha} - 1} \tag{5.9}$$

where T_n = amplifier noise temperature

T = diode temperature

G = amplifier gain

f_1 = signal frequency

f_2 = idler frequency

and α = term depending upon additional loading at idler frequency. (α is usually small enough to be negligible to a first-order approximation.)

The noise performance of frequency-converter amplifiers is determined by a similar equation.

It is possible to calculate the pump frequency f_3^* at which the parametric-amplifier noise temperature has its minimum value. This is given by

$$f_3^* = f_1 \left\{ 1 + \left(\frac{f_c}{f_1}\right)^2 \frac{\gamma^2}{1+\alpha} \right\}^{\frac{1}{2}} \qquad (5.10)$$

If α is small and $f_1/f_c \ll 1$, then this simplifies to

$$f_3^* = \gamma f_c \qquad (5.11)$$

and the optimum noise temperature for large gain becomes

$$T_n = T \frac{2f_1}{\gamma f_c} \qquad (5.12)$$

where T_n = amplifier noise temperature

and T = diode temperature

It will be realised, of course, that these equations represent the optimum performance of an amplifier, and that in practice this performance may not be easily attainable. In particular, it may not be possible to satisfy simultaneously the conditions for minimum noise temperature and maximum bandwidth. The equations do, however, give an indication of the noise performance which is likely to be approached, as a function of signal and pump frequencies and of diode quality.

68

From the equations for noise temperature it is possible to define a figure of merit M for the diode, where

$$M = \gamma f_c \qquad (5.13)$$

Hence
$$T_n = \frac{2f_1 T}{M} \qquad (5.14)$$

An alternative figure of merit is also often used. This is the 'dynamic quality factor', written as γQ or $\tilde{Q}$, and equal to M/f_1.

These, and various other figures of merit will give equivalent calculated performances for amplifiers, when applied to the same equivalent circuit for diode and amplifier. Noise-temperature calculations from the simple theory are likely to be misleading, however, unless allowance is made for the various parasitic elements in the amplifier structure, by replacing the values of γQ or M which apply to the diode with lower 'effective' values applicable to the amplifier as a whole.

However the terms are defined, it is apparent that for good amplifier performance a diode should combine a high cutoff frequency with a marked capacitance variation, even at low temperatures, in order to achieve low noise temperatures.

Eqn. 5.9 is written in terms of diode parameters, but for some purposes it may be more useful to consider the circuit behaviour of the amplifier. The noise temperature could be written

$$T_n = \frac{T_1 R_1}{R_g} + \frac{T_d R_s}{R_g} + \frac{f_1}{f_2} \frac{|R|}{R_g} \left(\frac{T_d R_s + T_2 R_2}{R_2 + R_s} \right) \qquad (5.15)$$

where $R =$ negative resistance

$R_g =$ generator resistance in signal circuit

$R_s =$ spreading resistance

and R_1 and $R_2 =$ losses (other than R_s) in signal and idler circuits.

At high gain, $|R| = R_g + R_s$

T_1, T_2 and T_d are the temperatures of R_1, R_2 and R_s, respectively. In practical amplifiers we can usually write $T_1 = T_2 = T_d = T$, the amplifier configuration being chosen so that, as far as possible, R_1 and R_2 are negligible compared with R_s. Then

$$T_n = T\left(\frac{R_g + R_s}{R_g}\frac{f_3}{f_2} - 1\right) \qquad (5.16)$$

For many applications, refrigeration is not necessary. Adequate low-noise performance can be achieved using simple circuits, with diodes of moderate cutoff frequency. C band parametric amplifiers are available from a number of manufacturers giving noise temperatures in the 300–400 °K region, with a pump frequency in X band. Such amplifiers have gains of the order of 20 dB, and bandwidths of 25–30 MHz.

The noise performance of such devices is, in general, quite adequate when used on systems operating at low angles of elevation, where the noise contribution from the atmosphere and the 'hot' Earth is significant.

If lower noise temperatures are required, it is necessary to use higher idler frequencies, and this is possible only if the cutoff frequency of the diodes is also increased. Examples of commercially available amplifiers of this type are shown in Plate 6. The high-frequency idler circuit is usually provided, in this type of amplifier, by making use of the series resonance of the diode to support this frequency, as shown schematically in Fig. 18.

When very high pump and idler frequencies are used, some of the assumptions implicit in the derivations of the above equations become less realistic. It has been tactily assumed that the amplifiers are strictly 3-frequency devices, and that power only flows at the signal, idler and pump frequencies. However, as f_3/f_1 is increased, the impedance of the pump circuit at the upper sideband frequency $f_3 + f_1$ will become more significant, and propagation at this unwanted frequency in the pump circuit will be possible.

Hughes and Pearson (1966a) have considered the contribution to the noise factor of an amplifier due to the presence of the upper sideband in the idler circuit. In certain amplifiers they find that this effect can seriously degrade the noise performance and may also reduce the available gain.

70

In some amplifier configurations, propagation of idler power in the pump circuit may also increase. Additional filter elements must be introduced into the amplifier to prevent the degradation of the amplifier performance by the unwanted power flow at these two frequencies.

Fig. 19 is a simplified diagram of this type of amplifier. The additional filter elements are indicated. At C band frequencies, such a device might have a noise temperature of about 100 °K, with a pump frequency of about 30–36 GHz.

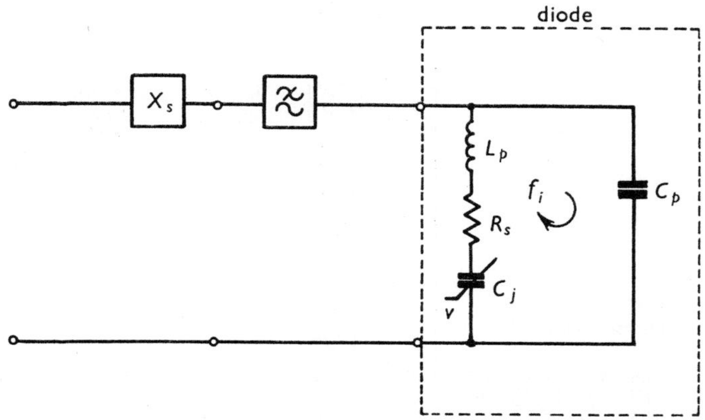

Fig. 18   Schematic circuit of parametric amplifier using
diode resonance to support idler

Although still lower noise temperatures may eventually be obtained, using unencapsulated diodes and much higher pump frequencies, for the present, very low noise temperature can only be achieved by reducing the temperature of the diode or of the whole amplifier.*

This requirement affects both the amplifier and diode design. The diode is required to maintain its various properties at temperatures of 77 °K, 25 °K or even below 4·2 °K. The amplifier design must take into account the requirements of economical cryogenic operation without introducing a large amount of additional attenuation (and hence thermal noise) into the amplifier input.

* For a review of cooled-amplifier design, see Uenohara (1967)

Many cooling systems have been used, ranging from relatively simple 'spot' cooling of the varactor alone (either by thermoelectric refrigerating modules, cold-gas streams or conduction along silver rods cooled externally by liquid air), to amplifiers entirely immersed in

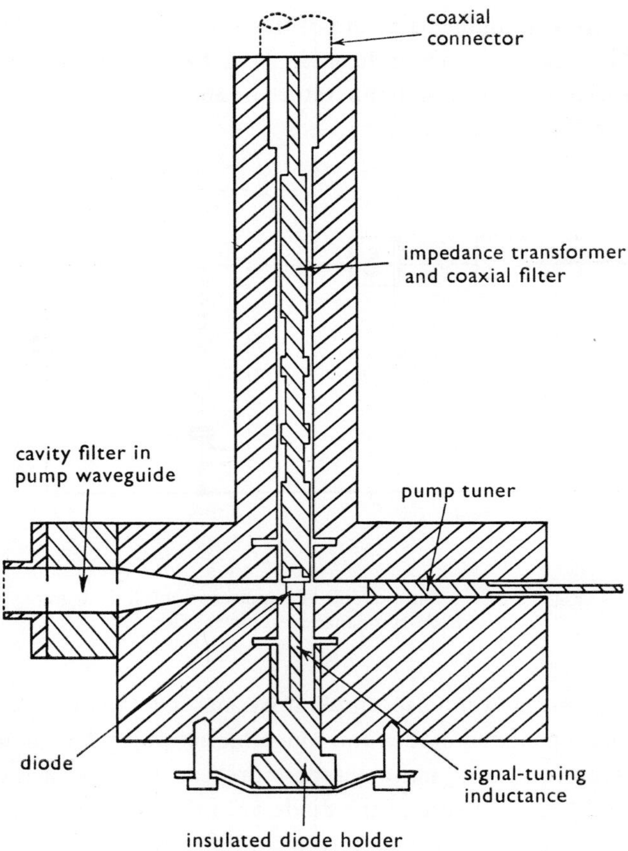

Fig. 19   Diagram of parametric amplifier

liquid nitrogen, and to amplifiers cooled with liquid helium and providing noise temperatures comparable with those previously attained only by masers. Most cooled amplifiers now under construction are, however, cooled to 16–20 °K by one of the small closed-cycle helium-gas refrigeration machines which are now commercially available, with liquid cooling restricted to a few particular applications.

When calculating the noise temperature of a cooled amplifier, allowance must be made for changes in diode parameters at low temperatures (see, for instance, Chakraborty and Coackley, 1967) as well as the inevitable additional sources of thermal noise introduced because of compromises between electrical and cryogenic requirements.

Even when such additional sources of noise have been accounted for, measurements on actual amplifiers often show a discrepancy between measured and calculated noise temperatures. For example, cooling an amplifier from ambient temperature to $77\,^{\circ}\mathrm{K}$ should reduce its noise temperature by a factor of four, whereas in practice the measured improvement is usually only by a factor of about three. The discrepancy is presumably due to a combination of small changes in diode characteristics, but no complete explanation is at present available.

The basic designs of actual amplifiers for different cooling arrangements are much the same, but the constructional details differ widely because of the cryogenic requirements.

The major problem is that the signal-input line must introduce a negligible amount of attenuation, but simultaneously provide a low heat leak into the refrigerated volume. Some of the arrangements which have been used are thin-wall stainless-steel or cupronickel-alloy waveguide with an inner plating of copper, waveguides of plastic or glass fibre with an inner layer of copper, silver or gold and complete mechanical heat breaks in standard waveguides with special choke flanges to provide microwave continuity. Similar arrangements are used for the pump and signal-output lines, although for these a somewhat greater microwave attenuation is tolerable. Of course, each line must be sealed from the atmosphere to avoid condensation of water, and, at lower temperatures, of oxygen and nitrogen, inside the amplifier.

When liquid nitrogen (or air) is used as a refrigerant, it must be excluded from the amplifier, usually by mounting the latter in a sealed container completely immersed in the liquid, as in Plate 7, or in a cold well projecting into the liquid, as in Fig. 20.

73

In either arrangement, thermal contact between the amplifier and the refrigerant is by radiation and metallic conduction. Additional conduction is provided by filling the enclosure with dry helium gas.

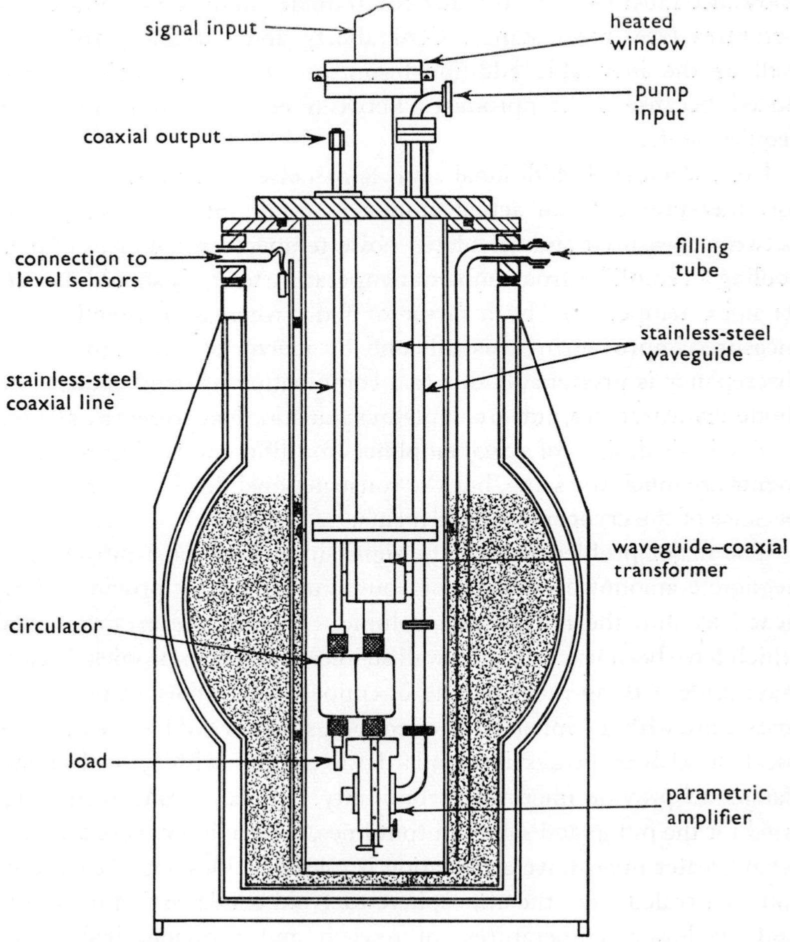

Fig. 20   Diagram of liquid-nitrogen-cooled parametric amplifier

Even though liquid nitrogen is cheap, frequent replenishment is inconvenient and heat leakage into the cold system must be minimised by choosing constructional materials of poor thermal conductivity, such as stainless steel, glass fibre etc., with thin copper or silver layers to ensure adequate electrical performance. A simple vacuum Dewar vessel is usually adequate to hold the liquid.

An example of a complete liquid-nitrogen-cooled parametric amplifier using a 'cold-well' design is shown in Fig. 20, and the complete equipment is shown in Plate 8.

For cooling to liquid-helium temperatures, the high cost and handling difficulties of the liquid make it essential for much greater care to be taken to ensure minimum leakage of heat into the structure, and multiple radiation shields, double Dewar vessels and other advanced cryogenic techniques are necessary. One advantage is that the electronic components can usually be immersed directly in the helium, which has a low permittivity and is not lossy. A cross-section through an amplifier of this type is shown in Fig. 21. The wide-bore Dewar vessel is particularly convenient for experimental work and the structure is designed to permit tilting through 45°, so that the amplifier can be mounted on a large movable aerial.

A liquid-helium-cooled amplifier is inevitably large, inconvenient to use, and of doubtful long-term reliability in an operational system. For these reasons there is an increasing tendency to rely on closed-cycle refrigerators to cool the amplifier, even though the noise temperature which can be achieved is somewhat higher than might be expected at liquid-helium temperature. A refrigerator giving about 1 W of cooling at 16–20 °K is much simpler and cheaper than a similar unit incorporating a Joules–Thomson expansion stage giving 0·5 W at 4·2 °K, and so the former type is much more likely to be used in operational systems. Because the available cooling power is small, the complete equipment must be designed for minimum heat leakage into the cold zone. The amplifier and refrigerator must be surrounded by a vacuum chamber containing multiple radiation shields.* Good thermal contact must, however, be maintained between the amplifier and the cold surfaces of the refrigerator.

Plate 9 shows a complete parametric-amplifier–refrigerator

* Several such amplifiers have been described. See, for example, some of the papers in 'Low temperature refrigeration' (Boston Technical Publishers, 1967). Some Russian cooled amplifiers have been described by Alfeyev (1966)

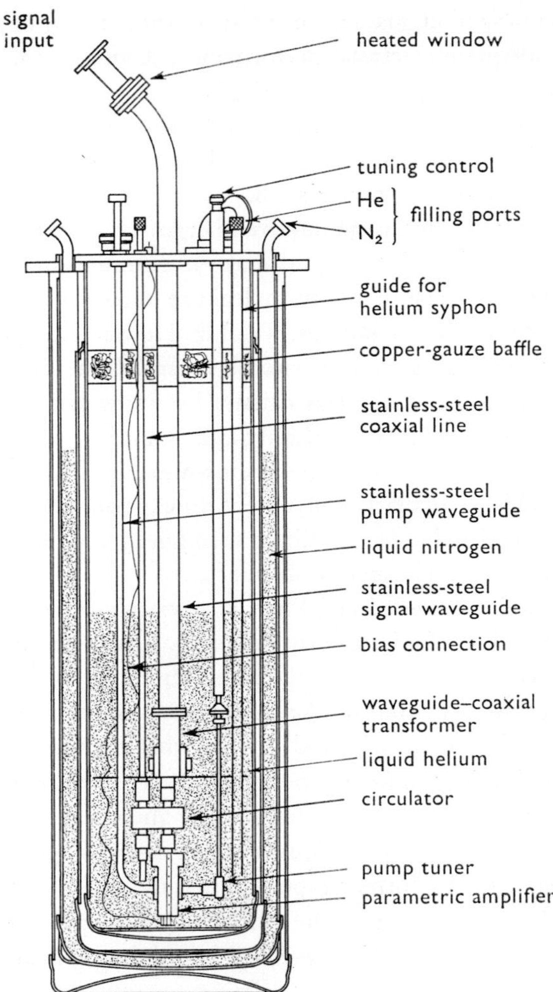

Fig. 21   Diagram of liquid-helium-cooled parametric amplifier

combination. Unlike liquid-cooled devices, amplifiers mounted on refrigerators can be used in any orientation.

A circulator is an essential part of any of these amplifiers, and is normally intimately associated with the varactor and the tuned circuits. Unfortunately, the ferrite materials normally used are unsuitable for use at low temperatures. Coolable circulators, using substituted yttrium–iron garnet, are at present available commercially for a limited number of frequency bands, but mostly the performance of the cooled device is, in a number of ways, inferior to that of the best available uncooled circulators. For use at frequencies for which cooled circulators are not available, a cooled amplifier would have to incorporate an uncooled circulator, and a significant deterioration in overall performance would be inevitable.

As the noise temperature of an amplifier is reduced, the importance of any shot noise increases. The theoretical contribution to the amplifier noise temperature from shot noise caused by a current flowing through the diode is given by Watson (1968) as

$$T_{shot} = \frac{e\,I}{8\pi^2 k R_g}\left\{\frac{1}{f_1^2 C_j^2} + \frac{1}{f_2^2 C_j^2}\left(\frac{\gamma f_c}{f_2}\right)^2\right\} \qquad (5.17)$$

which is usually 2–$10\,^\circ\mathrm{K}/\mu\mathrm{A}$ depending on the circuit and diode parameters. For uncooled gallium-arsenide varactors, experimental results are in reasonable agreement with such theoretical figures, although at high pump levels values in the region of 8–$10\,^\circ\mathrm{K}/\mu\mathrm{A}$ are measured (Watson, 1968). For silicon varactors, the noise contribution appears to be at least ten times greater (Josenhans, 1964). The discrepancies have been attributed to the movement of minority carriers having a lifetime greater than half a pump cycle.

5.3 Bandwidth

5.3.1 Single diode amplifiers

Much effort has been spent in devising ways of obtaining the broad bandwidth essential in some applications.

It can be shown (Aitchison, Davies and Payne, 1966 and 1968) that the gain–bandwidth product for a nondegenerate parametric amplifier with single-tuned signal and idler circuits can be expressed in terms of the diode parameters as

$$G^{\frac12}B = 2\left(\frac{f_c}{f_r^2} + \frac{f_2}{f_1\,\gamma^2 f_c}\right)^{-1} \qquad (5.18)$$

where $\qquad\qquad f_r =$ series self-resonant frequency

and $\qquad\qquad\;\; f_2 =$ idler frequency

The gain–bandwidth product is optimised by using the series resonance to support the idler current and it can be shown (Aitchison, 1967) that there is then an optimum idler frequency determined by the stray reactances of the diode encapsulation and given by

$$f_2 = \left(\frac{2f_1\gamma^2 f_c^2}{1 + C_s/C_j}\right)^{\frac{1}{3}}$$

(5.19)

The maximum gain–bandwidth product is then given by

$$G^{\frac{1}{2}}B = \frac{2}{3f_c}\left(\frac{2(\gamma f_c)^2 f_1}{1 + C_s/C_j}\right)^{\frac{2}{3}}$$

(5.20)

For a typical diode with an f_1 of 4 GHz, $G^{\frac{1}{2}}B$ could be somewhat greater than 1 GHz.

This performance is inadequate for, for example, a high-capacity satellite-system earth station. The usual technique for increasing the bandwidth is reactance compensation, i.e. the addition of one or more additional tuned elements to the signal circuit. Theoretical analysis, supported by experiment, shows that the gain–bandwidth product becomes of the form $G^{\frac{1}{4}}B = $ constant for one additional circuit and $G^{\frac{1}{6}}B = $ constant for two additional circuits. For a maximally flat design, the limiting bandwidth should be given by $B\log G = $ constant, but in practice the use of more than two or three compensating elements leads to practical difficulties in tuning the device. Plate 10 shows the simple amplifier of Fig. 19 to which two sets of coaxial stubs have been added to increase the bandwidth.

5.3.2 Double-diode amplifiers

DeJager has carried out an analysis of the bandwidth to be expected in a parametric amplifier, and this includes the effects of both parasitic elements in the diode package and isolating filters between the signal and idler circuits (DeJager, 1964). By introducing two 'slope factors' d_1 and d_2 for the signal and idler circuits, DeJager derives a modified equation for the gain–bandwidth product. The factors d_1 and d_2 represent the reactance slope of the signal and idler circuits near ω_1 and ω_2, respectively, compared with the reactance slope of the mean capacitance C.

According to DeJager, the optimum value of the gain–bandwidth product will be proportional to $(d_1 d_2)^{-\frac{1}{2}}$, and there appears to be minimum $d_1 d_2$ of four. Although it has not been shown that the limitation $d_1 d_2 \geqslant 4$ *must* apply to *all* amplifiers, no single-diode configuration has yet been shown to yield a smaller $d_1 d_2$. The limitation does not apply however to degenerate amplifiers, nor to amplifiers using two identical diodes in a balanced configuration. For these, it is theoretically possible for d_1 and d_2 to equal unity, and for the original bandwidth calculations to apply. It follows that, under idealised conditions, the limiting bandwidth of a balanced-diode amplifier might be expected to be twice that of a single-diode amplifier. Edrich (1966) has shown that under realistic conditions this ideal factor of two cannot be approached very closely. Nevertheless, the double-diode approach has a number of advantages and has led to several practical amplifier designs.

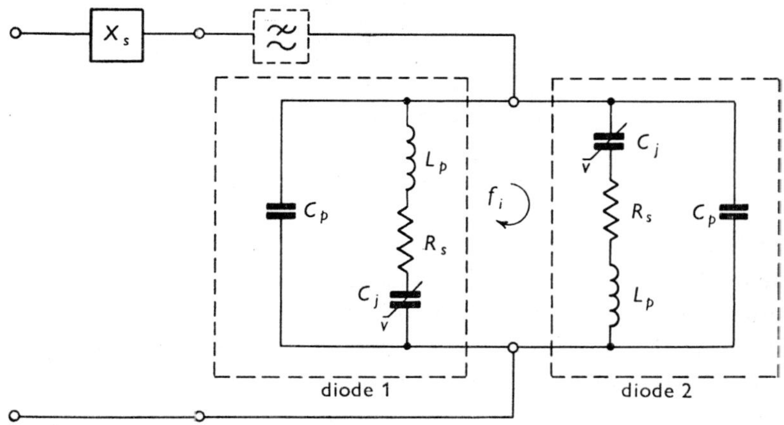

Fig. 22   Schematic circuit of double-diode parametric amplifier

The balanced nature of the amplifier can give a very broad idler bandwidth and at the same time ensure good decoupling between the signal and idler circuits. Identical diodes are used, mounted so that the individual idler voltages are developed across the diodes in anti-phase. The net idler voltage thus cancels out. As no net idler voltage is generated across the capsule capacitances, these do not form part of the effective idler circuit. To achieve this balance, the pump frequency must be chosen so that the idler frequency is determined by L_s and C_j.

The two diodes and the short connecting loop then form the idler circuit. The equivalent circuit can be drawn as in Fig. 22.

A number of different arrangements have been proposed to realise this concept. Kliphuis (1961) used two diodes mounted 'back-to-back' across reduced-height pump waveguide, with the signal introduced at their common point. Pearson and Lunt (1964) reduced the height of the pump waveguide still further and mounted two diodes side by side across the guide. In a later version (Hughes and Pearson, 1966b) the two diodes were connected within a single encapsulation (see Fig. 23). When the two diodes are side by side, the pump field can be coupled to

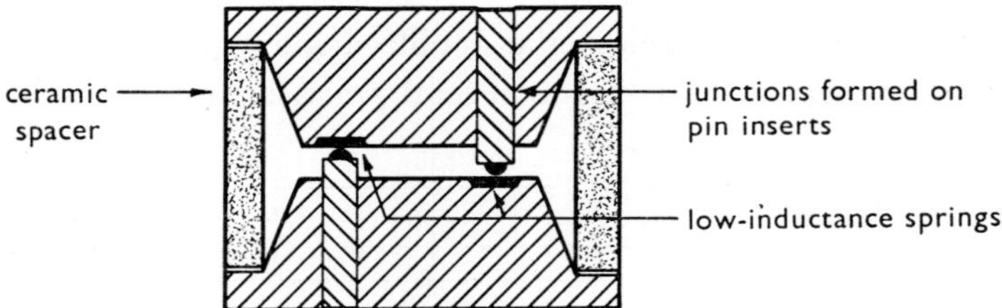

Fig. 23   Double-diode encapsulation
(Courtesy Ferranti Ltd.)

the diodes through either the E vector or the H vector. For the former, the diodes are situated as close together as possible and mounted in antiparallel to ensure that the two idler voltages cancel out. For the latter, the diodes are mounted in parallel to form a coupling loop, and as in the back-to-back arrangement, it is comparatively easy to monitor the total current flowing through the two diodes, and, if necessary, to apply external bias. A simple double-diode amplifier is shown in Fig. 24. In a further development, the two varactors are separated by a microstrip line half an idler wavelength long (Aitchison, 1967), as shown in Plate 11.

For some applications, a very wide tuning range is more important than a wide instantaneous bandwidth. Because of the absence of idler filters, the double-diode type of amplifier is particularly suited to this type of application. The idler frequency can be varied, within limits set by the onset of significant current flow through the diodes, by varying the applied bias voltage. The balanced arrangement has enabled very compact amplifiers to be designed, and an example is shown in Plate 12.

It should be noted that the noise-temperature expression (eqn. 5.9) is equally applicable to single- and balanced-diode amplifiers, and for equal idler frequencies, equal noise temperatures should be obtained. However, in general, even when the same diodes are used in the two types of amplifier, the optimum idler frequencies are unlikely to be the same, and therefore different noise temperatures must be expected.

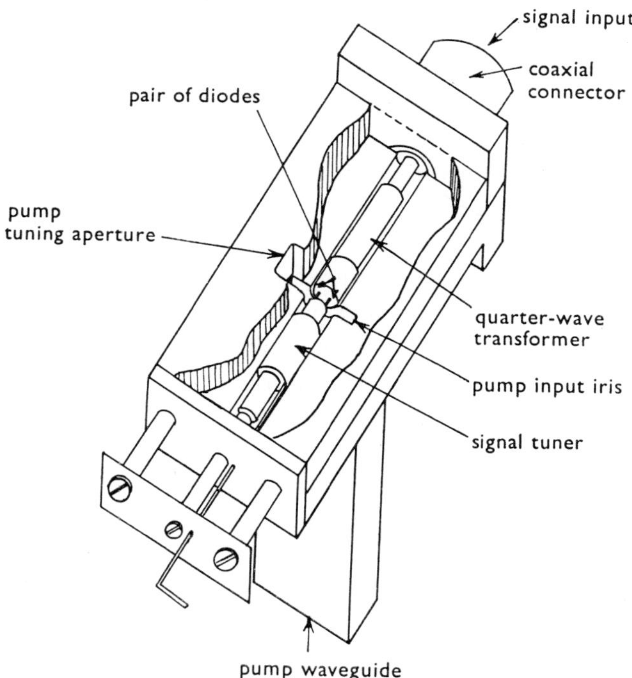

Fig. 24   Diagram of simple double-diode parametric amplifier

5.4 Dynamic range

Although the low noise temperature and the attainable bandwidth are probably the most important characteristics of the parametric amplifier, other properties are important in many applications. For example, the behaviour at high signal levels is important.

The dynamic range is limited by the basic noise at low levels, and by gain compression at higher levels. Unlike the gain-compression phenomena in a maser, in a parametric amplifier nonlinear effects increase very rapidly near the gain-compression point. Below the point at which gain compression becomes significant, third-order intermodulation output power is given by

$$P(2f_1 - f_2) = 2P(f_1) + P(f_2) + C \qquad (5.21)$$

where C is between -5 and $+28$ depending upon operating frequency and amplifier design, and all values are expressed in decibels with respect to 1 mW.

This may be contrasted with a maser, where C is of the order of -92 dBm (see eqn. 4.12). Nevertheless, the maser has exceptional intermodulation performance, and, for almost all likely low-noise applications, the parametric-amplifier performance should be quite acceptable.

At high signal levels, gain compression will occur. The overall gain of the amplifier depends upon the magnitude of the negative resistance generated by the application of pump power. The magnitude of the negative resistance depends upon the voltage across the diode at the pump frequency. If this voltage drops, only slightly, under conditions of high gain, the gain drops appreciably. If a constant pump power is used, increasing the signal level will increase the admittance seen by the pump source and the voltage will consequently fall. Hence the gain of the amplifier decreases as the signal level increases, i.e. saturation occurs.

The effect can be delayed by increasing the pump power to keep the pump voltage constant. If this is done, second-order effects become important and the simple equation for the variation of capacitance no longer applies. The result is that additional capacitances are introduced into the effective equivalent circuit, the circuits are detuned and the gain and noise temperature deteriorate (Brenner, 1967).

Since the available output power is shared between the idler and signal frequencies, the gain limitation at signal frequency will occur at a lower output power level for amplifiers with high idler-frequency/signal-frequency ratios than for those with a low ratio, and so very-low-noise amplifiers will tend to have reduced saturation-power levels. Uenohara and Elward (1964) quote, as a rule of thumb, that gain compression occurs when the total power emitted from the junction at idler and signal frequencies is about 10 dB below the level of pump power at the diode. One particular refrigerated broadband amplifier has been quoted as having 0·5 dB gain compression at an output-signal level of about -40 dBm.

Greater saturation will occur if the signal level is further increased, although of course, in a real system, saturation will also occur in other elements in the system and may mask the behaviour of the parametric amplifier. When the overload is removed, and provided that the diode has not been damaged by the overload, the amplifier gain should be restored extremely rapidly, with time constant determined by the Q factors of the circuits involved.

Little information is available on the ability of the varactor to withstand high-level overload without burnout. This is usually a problem with small-area point-contact mixer diodes which have burnout levels of less than an erg. However, tests have shown that both silicon and gallium-arsenide diffused mesa varactors will usually withstand levels in excess of 20 ergs. This suggests that these devices are relatively rugged, but further tests are still necessary.

5.5 Comparison between the maser and the cooled parametric amplifier

These two types of amplifier are often used for similar applications. The comparison in Table 2 indicates the most important differences between them.

Table 2

Property	Parametric amplifier	Maser
Dependence on:		
(a) pump amplitude	critical	usually saturated
(b) pump f.m.	sensitive	not affected
(c) pump frequency	sensitive	not very critical
(d) magnetic field	no	yes
(e) bias	yes	no
(f) physical temperature	relatively insensitive	critical
Nonreciprocity from	external circulator	internal isolators
Bandwidth	wide	moderate
Delay distortion	low	moderate
Recovery time	microseconds	milliseconds

5.6 Other amplifier configurations

5.6.1 Degenerate parametric amplifiers

In the degenerate parametric amplifier, the pump frequency is approximately twice the signal frequency. The signal and idler pass-bands overlap, and instead of the idler circuit being terminated inside the amplifier, it is effectively terminated in the input of the amplifier. Usually this means that the idler is terminated in the sky, via the aerial.

Such an amplifier has a theoretical effective noise temperature of only half the optimum value for the same diode in a nondegenerate amplifier, with the additional advantages of a low pump frequency and a wider bandwidth. The absence of a separate idler circuit makes such an amplifier much simpler than the corresponding nondegenerate version. However, for most applications, these advantages are of little value compared with the major disadvantage that, because a signal entering the aerial appears in both idler and signal responses of the amplifier, the output contains both the signal and its image. When coherent communication signals are involved, this double response is unacceptable, although where the 'signal' takes the form of noise, as in some radioastronomy applications, the degenerate amplifier may be of considerable value. Other applications include measurements of varactor performance.

5.6.2 Multiple-idler amplifiers

Amplifiers can be designed with two or more idler frequencies, and it is then possible to use a pump frequency lower than the signal frequency. This is of particular value for very high signal frequencies. In the 2-idler case, the usual restriction

$$f_{idler} = f_{pump} - f_{signal} > 0$$

is replaced by

$$f_{I1} = f_{signal} - f_{pump} \quad \text{(first idler)} \tag{5.22}$$

$$f_{I2} = (f_{I1} - f_{pump})$$

$$= 2f_{pump} - f_{signal} > 0 \quad \text{(second idler)} \tag{5.23}$$

Therefore
$$f_{pump} > \tfrac{1}{2} f_{signal}$$

A 35 GHz amplifier has been built, using a 29 GHz pump, so that the two idler frequencies are 6 and 23 GHz (Vilcans and Ginsberg, 1964).

If f_{pump} is chosen as $\frac{2}{3}f_{signal}$, then the two idler frequencies will both be equal to $\frac{1}{3}f_{signal}$. Amplifiers of this type, operating at 2 GHz, have recently been described (Pommereit, 1967).

5.6.3 Travelling-wave parametric amplifiers

Most experimental work and theoretical analysis have been directed towards amplifiers which use either one or two varactor diodes. Just as the travelling-wave maser is related to the cavity maser, so attempts have been made to develop travelling-wave parametric amplifiers as alternatives to the cavity type. The travelling-wave parametric amplifier has the additional problem compared with the maser that it is necessary to provide the correct termination at idler frequency. It is therefore necessary for the travelling-wave structure to propagate both signal and idler, and in some arrangements pump frequency as well.

Basically, the amplifiers usually take the form of a transmission line periodically loaded with varactor diodes, and, for amplification to occur, the correct phase relationships must be satisfied. For optimum gain, the phase coefficients for the line at the signal, idler and pump frequencies should be satisfied by the equation

$$\beta_s + \beta_i = \beta_p + \frac{2m\pi}{l} \qquad (5.24)$$

where β_s, β_i, β_p are phase coefficients, l is the diode separation distance, and m is an integer.

It is obviously difficult in practice to satisfy this relationship, and some experimental amplifiers have provided separate pump feeds for each diode.

For frequencies below 1·5 GHz, experimental amplifiers have usually taken the form of stripline with the diodes between the central and outer conductors. For frequencies above 1·5 GHz, the diodes are usually mounted across rectangular waveguide or a series of coupled cavities. Alternatively, a helix structure has been used with the diodes bridging individual pairs of turns. As well as the normal forward-wave amplification, the structure might be designed so that the signal and idler waves propagate in opposite directions.

Considering the amplifiers described in the literature, it would appear that although a greater bandwidth should be attainable with the travelling-wave parametric amplifier compared with the more conventional designs, in practice it would appear that bandwidths of the order of 10–15% can only be achieved with gains of less than 15 dB and by the use of high pump powers and a large number of diodes. The noise performance and stability have been degraded by spurious backward waves. It would appear that, at present, the performance of travelling-wave parametric amplifiers can be exceeded by both single- and double-diode cavity amplifiers.

It is presumably because of the difficulties of providing a considerable number of identical diodes, the experimental difficulties of providing the correct phase characteristics at the three frequencies and the relatively high pump powers needed, that travelling-wave parametric amplifiers have not so far been developed beyond the experimental stage, and it does not seem probable that such development is likely in the near future. There is, of course, always the possibility that a distributed varactor might be produced, replacing individual diodes by a continuous strip of active material within a suitable transmission line.

5.6.4 Nonreciprocal or unilateral parametric amplifiers

As indicated in the previous section, multidiode travelling-wave parametric amplifiers have not proved very successful. There seems to be more likelihood of success with simplified versions in which only two diodes are used. A number of experimental amplifiers have been described, mainly for much lower frequencies than those considered above, but a number of proposals for nonreciprocal microwave amplifiers have been made.* With only two diodes, it is clearly much easier to satisfy the phase relationship of eqn. 5.24.

One possible amplifier configuration consists of two identical circuits, each consisting of signal-tuning reactances, an idler-rejection

* See, for example, Baldwin (1961), Maurer (1963), Thompson (1961) and Jones and Hyde (1967)

86

filter, a varactor and an idler-tuning reactance. The two circuits would be connected to two ports of a 3 dB hybrid, and would also be connected to a common idler load. The two varactors would be pumped from a common source with an extra length of line to introduce a 90° phase shift between the pump voltages at the two diodes.

Such an arrangement would provide gain between the third and fourth ports of the hybrid, but attenuation in the reverse direction, and could be used to provide amplification at frequencies at which circulators are not available.

5.7 The semiconductor varactor diode

The requirements for a variable-capacitance diode have been discussed in the preceding sections of this chapter. In the following sections, the problems of meeting these requirements in practical devices are considered.

It has been shown that, for a low noise temperature, the main requirement in a varactor diode is a good γf_c characteristic, in which an enhanced γ will also lead to improved bandwidth. The design of microwave varactor diodes to provide these parameters must take into account the semiconductor material, the junction element and the device encapsulation.

5.7.1 Semiconductor material

Varactor diodes have been made from a wide range of semiconductors, and it can be shown that the cutoff frequency, which we have defined by

$$f_c = \frac{1}{2\pi R_s C_j} \tag{5.8}$$

can be related to the properties of the material (Meredith and Warner, 1963) as follows:

$$f_c \propto \mu \epsilon^{-\frac{1}{2}} \tag{5.25}$$

So the perfect semiconductor should possess a low relative permittivity ϵ and a high carrier mobility μ. On comparing the various possible materials on this basis, with parameters corresponding to similar diode characteristics at their normal operating temperatures, indium antimonide with $\mu = 100\,000$ cm^2/V s and $\epsilon = 17$ would appear to be outstanding. Varactor diodes made of this material have been reported to have reasonable cutoff frequencies and have provided low-noise amplification (Allen *et al.*, 1963). However, as this material has an energy gap of only 0·18 eV, it requires cooling to below 77 °K to reduce the number of intrinsic carriers to an insignificant level, and this considerably restricts the usefulness of this type of diode. Of the materials which can operate over a wide range of temperatures, gallium arsenide ($\mu = 5000$ cm^2/V s, $\epsilon = 12\cdot5$) and silicon ($\mu = 1000$ cm^2/V s, $\epsilon = 11\cdot8$) appear to be the most promising. The energy gap of silicon (1·08 eV) allows operation up to 175 °C, but at low temperatures the carriers freeze out, degrading the performance, although by using heavily compensated material satisfactory characteristics are obtained at 77 °K (Fink and Rulison, 1964 and Benny, 1965). Diodes made of gallium arsenide, having an energy gap of 1·4 eV, also operate at high temperatures but have the advantage of retaining good characteristics down to liquid-helium temperatures for very-low-noise applications. The only changes observed in the diode characteristics have been associated with the expected variation in the diode contact potential. This ability to operate over a wide temperature range combined with the high cutoff frequency resulting from the high carrier mobility of gallium arsenide make it probably the most suitable varactor material for low-noise amplifiers.

The other factor in the figure-of-merit equation (eqn. 5.13) is the capacitance-variation coefficient, which also depends on the choice of semiconductor material. This can affect the capacitance/voltage law of the junction and determine the bias that can be applied to the junction. In general, the III–V group of compounds (indium antimonide, gallium arsenide etc.) have anomalous diffusion characteristics giving an abrupt junction with $C \propto V^{-\frac{1}{2}}$ as compared with $V^{-\frac{1}{3}}$ for the graded junction in silicon. The portion of this capacitance variation that can be used in the varactor diode is determined by the extent to which the device can be driven without incurring excessive noise. In the reverse direction, a breakdown voltage of about 2–6 V is adequate, but in the forward direction the voltage at which conduction takes place is related to the energy gap of the semiconductor material (and is about 0·1 V for InSb, 0·5 V for Si and 0·8 V for GaAs), and this can restrict the effective capacitance variation.

5.7.2 Junction element

As the electrical performance of a varactor diode is defined by the cutoff frequency, the capacitance variation and the breakdown voltage, the design of the junction element involves producing the optimum relationship between the series resistance and capacitance for a given breakdown voltage. Breakdown voltage is a function of the semi-conductor resistivity, and, for a parametric amplifier where only a low breakdown voltage is required, it is always advantageous to use the lowest possible resistivity in order to minimise the series resistance. By epitaxial-deposition techniques, it is possible to grow a thin high-resistivity layer on a heavily doped substrate. With material of this type the junction can be formed in the thin high-resistivity layer, and the breakdown voltage and other parameters associated with that doping level retained, but the low resistivity of the substrate reduces the bulk resistance of the device. The resistance of the diode includes not only the resistive components of the junction and bulk semicon-ductor, which can be calculated from the dimensions and resistivity, but also the contact resistances and skin effects at microwave fre-quencies. Although these are the factors that limit varactor diode performance in practice, they are not fully understood.

Ideally, the junction capacitance should be larger than any fixed stray capacitance of the encapsulation, and yet less than about 1 pF for a suitable microwave impedance. However, experiments indicate that the cutoff frequency is inversely proportional to the junction radius, as $C_j \propto r^2$ and $R_s \propto r^{-1}$. Therefore some compromise is necessary, and in practice the optimum performance is obtained with a junction capacitance in the region of 0·1–0·5 pF.

5.7.3 Junction fabrication and performance

There are many methods of forming a *p–n* junction and most of these can be applied to the fabrication of a varactor diode. Although some varactor diodes have been made by alloying or pulse bonding, the variability of this basic process leads to large tolerances in manufacture which make it difficult to attain a narrow base width for a low series resistance and outweighs the advantage of an abrupt junction with good capacitance variation.

The majority of devices are at present made by diffusion processes, which enable a higher order of control to be obtained over the junction geometry. Therefore epitaxial layers of a few micrometres thickness can be employed to realise a low series resistance and a reasonable breakdown voltage. Most of these junctions are reduced to the requisite area (or junction capacitance) by a mesa etching technique as shown in Fig. 25 a. With the current improvement in planar techniques in which an oxide mask is used to define the diffused area, it is now possible to create the small junction structures required for varactor diodes, and these processes will certainly be used to optimise the junction structure and geometry. This ability to fabricate small complex geometries has made it possible to attempt to produce multiple junctions for enhanced capacitance variation at high cutoff frequency and annular base contacts for a low microwave series resistance. Diodes have been made from indium antimonide, silicon, and gallium arsenide, but indium-antimonide diodes are unstable in practice and little work is now being done on this material for varactor diodes. There is a wide range of diffused-junction gallium-arsenide and silicon diodes available, and in general, the gallium-arsenide devices are employed for their low-noise properties whereas the silicon types can provide better bandwidth characteristics.

A structure consisting of a metal layer deposited upon an oxide film grown on a semiconductor substrate is known as a metal–oxide–semiconductor (m.o.s.) diode. Such a diode has a voltage-variable capacitance and is therefore potentially useful as a varactor diode (Howson *et al.*, 1965). It could have several advantages compared with a junction diode when used in a parametric amplifier, such as negligible current giving low shot noise and a large capacitance variation. However, in practice it has proved difficult to obtain a low series resistance or to control the threshold voltage.

The earliest metal–semiconductor diodes were formed by a point-contact method and, as this is best suited to producing very small junction areas, it has been mainly applied at higher frequencies (De Loach, 1964). Although high cutoff frequencies can be attained (Foxell and Wilson, 1965), these diodes exhibit a small effective-capacitance variation, with the overall result that they have only a

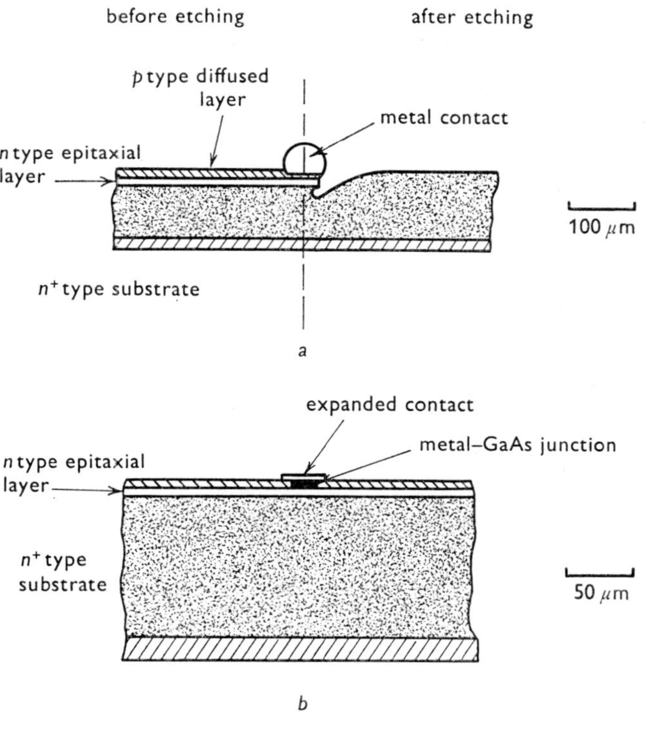

Fig. 25   Varactor diodes
a   Diffused gallium-arsenide mesa diode
b   Metal–gallium-arsenide diode

moderate figure of merit. However, this type of junction is formed by the metal-whisker contact on the semiconductor, and more recently the great advances in planar and oxide-masking techniques have made it possible to evaporate metal contacts directly onto the semiconductor. This Schottky-barrier, or hot-carrier, diode has a true abrupt junction with a consequently good capacitance variation, and a wide range of controlled electrical characteristics can be obtained with different evaporated metal contacts. As this junction does not penetrate the surface of the semiconductor, a very thin (< 1 μm) epitaxial layer can be used to give a lower series resistance than other junction devices. The construction of this type of device is shown in Fig. 25 b. At present, these diodes are being rapidly developed with characteristics comparable to the earlier diffused devices, and there seems little doubt about the long-term importance of this form of diode (Foxell and Summers, 1965).

The performance of the various types of varactor diode are summarised in Table 3.

Table 3

Type	Semi-conductor	Typical * characteristics f_{co}, GHz	γ	Advantages	Disadvantages
Alloyed	InSb, Ge, Si, GaAs	50	$< 0\cdot2$	good γ	difficult to fabricate in a controllable manner
Diffused	InSb	150–400	$<0\cdot1$	high f_{co}, can be used at 4 °K	unstable, cannot be used above 77 °K
	Si	100–200	$0\cdot2$	reasonable f_{co}, good γ	cannot be used much below 77 °K
	GaAs	100–500	$0\cdot15$	good f_{co}, reasonable γ can be used at 4 °K	—
Point contact	GaAs	100–750	$<0\cdot1$	good f_{co}	poor γ
Metal–semi-conductor	Si	100–200	$0\cdot2$	as diffused type but with possibility of higher f_{co}	—
	GaAs	300–500	$0\cdot15$		—
M.O.S.	Si	100	—	potentially high γ	poor f_{co}

* As measured by transmission technique: f_{co} at zero bias; γ as defined by eqn. 5.3 between the limits $I_f = 1$ μA and $V_R = -1$ V

Brenner and Eisele (1966) have measured changes in γ which take place when diodes are pumped hard. Their results show a noticeable difference between the behaviour of silicon-epitaxial and various types of GaAs diodes. The silicon diodes exhibit a large increase in γ with increasing diode forward current, but little is gained by changing the bias voltage. GaAs diodes show little increase in γ with increasing current, but there is a steady increase with increasing negative bias when the forward current is fixed.

It was concluded from the data that the gain–bandwidth products of silicon epitaxial diodes can be improved at the expense of increased shot noise, which must be expected at larger forward currents, by pumping slightly into the forward conduction region, whereas little can be gained by using the same technique with GaAs diodes.

5.7.4 Encapsulation

Basically, the encapsulation can limit the performance of a varactor diode by having (*a*) a large stray capacitance in relation to the junction capacitance, (*b*) an excessive series inductance giving low series- and parallel-resonant frequencies and (*c*) poor thermal properties. Most varactor cases consist of two metal connections carrying the diode element separated by a ceramic insulator (see, for example, Plate 13). In general, reducing the size of the encapsulation by bringing the metal parts closer together will give a lower series inductance, but, owing to limitations on the minimum size of ceramic insulators that can be fabricated, this will tend to increase the stray capacitance, and therefore some compromise must be made between the stray capacitance and inductance. Apart from the inductance of the metal-case components, there is also a contribution from the internal connections.

In general, the stray capacitance of a typical alumina ceramic encapsulation is about 0·1–0·3 pF, and some effort is being devoted to reducing this by the possible use of quartz or beryllium oxide as an insulator.

Most varactor diodes for parametric amplifiers have series-resonant frequencies of about 9 GHz (Fig. 26*a*), and the problem of achieving higher frequencies for the idler has been approached by combining two diodes within one case (Benny and Pearson, 1965) with series resonant frequencies of about 20 GHz. However, the alternative method of reducing the overall size of the encapsulation to a 'micro-pill' case (Fig. 26 b) to give a lower inductance has provided series-resonant frequencies in excess of 30 GHz (Foxell and Summers, 1965). It is likely that with the introduction of the Schottky-barrier diode a further improvement in series-resonant frequency will be obtained on dispensing completely with the encapsulation and mounting the diode chip directly into the circuit (Fig. 26 c).

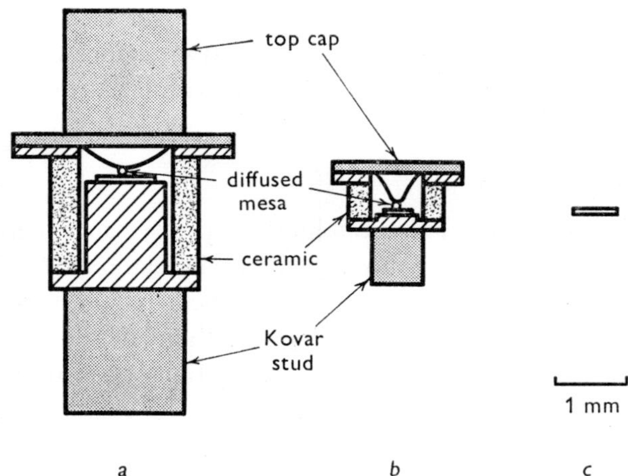

Fig. 26   Comparison of varactor diodes
a   Ceramic-pill case with series-resonant frequency of about 9 GHz
b   Ceramic micropill case with series-resonant frequency of about 30 GHz
c   Unencapsulated planar semiconductor chip

Good thermal dissipation is necessary to avoid the generation of additional thermal noise by the spreading resistance when the diode is heated by the pump power (Garbrecht, 1965).

5.7.5   Measurement

The measurement of diode characteristics in the microwave region is difficult. Two basic techniques have been evolved: (*a*) the reflection method of Houlding and Harrison which only gives information on the semiconductor region, and (*b*) the transmission method of De Loach and Wilson which determines the complete equivalent circuit. The use of the reflection technique is uncertain owing to losses in the measurement system which lead to low values of f_{co} and high values of γ. Recently it has been shown (Doherty, 1966) that, with a high-precision line, good agreement can be obtained with the transmission method.

5.8 Future trends

The varactor diodes at present available are capable of providing very low noise temperatures in parametric amplifiers when cooled to liquid-helium temperature, and therefore the basic aim of varactor development must be to realise a device capable of giving a noise temperature of 25–50 °K at room temperature. Compared with current values of γf_{co} of about 50 GHz, this requirement would correspond to a figure of the order of 100 GHz.

The recent improvements in both gallium-arsenide and silicon varactors have been due to the use of epitaxial techniques, and now planar rather than mesa processes are being applied to produce the small junctions required. However, the new form of diode based on a metal–semiconductor junction (Schottky barrier) is almost certain to replace the existing diffused varactors. In both Si and GaAs this has the advantage of a true abrupt junction with a good capacitance variation, and, as this junction is produced by an evaporated-metal contact which does not penetrate the semiconductor, a very thin epitaxial layer can be employed to give a low series resistance. Therefore the exploitation of this device will depend on the quality of the epitaxial material and for GaAs this is likely to become a critical factor.

The encapsulation is also an important aspect of varactor performance in that the resonant frequency can limit the idler and hence reduce the bandwidth. In the future, the solution will probably lie in integrating the diode into the amplifier as this will considerably reduce the parasitics associated with the encapsulation, and it is significant that planar diodes are now being fabricated which should enable this to be accomplished.

The use of unencapsulated diodes should enable parametric-amplifier operation to be extended satisfactorily into the millimetre-wave region, and should also enable considerable improvements to be made in the noise and bandwidths at lower frequencies.

6 MICROWAVE-MIXER AND DETECTOR DIODES

6.1 Introduction

This section is concerned with diode rectifiers for use in super-heterodyne and crystal video microwave receivers without consideration of r.f. preamplification.

Heterodyne reception provides conversion of the carrier frequency of a signal to a new value. This is accomplished by means of a local oscillator and a nonlinear element. The local oscillator and the signal are coupled into the nonlinear device and generate (among other frequencies) the difference frequency between the signal and local-oscillator frequencies, normally called the intermediate frequency. The amplitude of the signal produced at this frequency is proportional to the input-signal amplitude and is substantially independent of the amplitude of the local-oscillator output. The circuit which contains the nonlinear element, the means for introducing the signal and local oscillator, and which provides the output terminals for the intermediate frequency, is usually called the mixer, and the process is referred to as mixing or frequency conversion. The nonlinear element is usually a rectifying semiconducting device, known as a mixer diode. The intermediate frequency voltage is amplified, and the complete circuit is referred to as a superheterodyne receiver.

Recent developments in semiconductor-diode technology have enabled considerable advances to be made in mixer and detector noise performance, and these are discussed in the following sections. A detailed discussion of mixer and detector mounts will not be included, as adequate descriptions are available elsewhere.

For radar applications (i.e. transmit–receive systems), an additional consideration is the need to protect the receiver from the high-power transmitter. The portion of the radar system that connects the transmitter and receiver to a common antenna and directs both transmitted and received powers to the proper destination is known as the duplexer. An example of a simple X band transmit–receive system is illustrated in Plate 14.

'Video crystal' is the term usually used for a diode rectifier that is used as a square-law detector of microwave pulses. The video output voltage of the detector is amplified, and the circuit is usually referred to as a crystal-video or detector-video receiver. The sensitivity of such a receiver is low compared with that of a superheterodyne receiver. For example, the video receiver is capable of detecting signals of about -50 dBm compared with about -100 dBm for a superheterodyne receiver with a comparable bandwidth (about 1 MHz). However, this is not prohibitive in many cases, and the advantage of simplicity, small size and wide bandwidth makes this an attractive system for applications where high sensitivity is not demanded. An example of a simple X band crystal-video receiver is illustrated in Plate 15.

Although the overall noise factors of superheterodyne receivers and detectors can be defined similarly to those of low-noise amplifiers, the equations are often written slightly differently. For two networks in tandem, eqn. 1.6 may be written in terms of noise factor, rather than noise temperature, as

$$F_{(1,2)} = F_{(1)} + \frac{F_{(2)} - 1}{G_{(1)}}$$

where $\qquad F_{(1)}$ = noise factor of network (1)

$\qquad F_{(2)}$ = noise factor of network (2)

and $\qquad G_{(1)}$ = available power gain of network (1)

The bandwidth of network (2) is assumed to be less than that of network (1).

If $\qquad F_{(1)} = \dfrac{t_{(1)}}{G_{(1)}}$

then $\qquad F_{(1,2)} = \dfrac{t_{(1)} + F_{(2)} - 1}{G_{(1)}}$ (6.1)

and an equation of this form is generally used in the discussion of mixer performance.

98

6.2 Mixer diodes

6.2.1 Superheterodyne-receiver performance

For a superheterodyne receiver, $t_{(1)}$ is replaced in eqn. 6.1 by N_r and $1/G_{(1)}$ by L_c, so that, following Torrey and Whitmer (1948), and assuming that any local-oscillator noise is suppressed,

$$F_0 = L_c(F_{i.f.} + N_r - 1) \tag{6.2}$$

where

L_c = conversion loss of the mixer diode

N_r = noise-temperature ratio of the mixer diode (defined as the ratio of the available noise-power output of the mixer to that of an equivalent resistor at $290\,°\text{K}$)

and $F_{i.f.}$ = noise factor of following i.f. amplifier

6.2.2 Image-matched conditions

The double-channel or image-matched input circuit is favoured for most radar-receiver applications, and the noise-factor performance of mixer diodes is measured and specified under these conditions (normally called the broadband case, i.e. when the same load admittance is presented to the image and signal channels). The theoretical limit for the diode conversion loss set by this condition is 3 dB, and so the theoretical limit for the receiver performance, assuming $N_r = 1$, is given by

$$F_0 = 2F_{i.f.}$$

e.g. 4·5 dB for an i.f. amplifier with a noise factor of 1·5 dB.

So, as the diode performance approaches 3 dB, the performance of the i.f. amplifier has increased significance in relation to the receiver performance.

6.2.3 Image-tuned circuits

The image-matched conditions are not, however, those for minimum conversion loss, for it may be shown that conversion loss is dependent on the image termination (Torrey and Whitmer, 1948). Briefly, conversion of signal power to the intermediate frequency entails the production of power at the image frequency. In the image-matched conditions this power is dissipated in the image termination and represents a serious drop in conversion efficiency. However, if the power at the image frequency is reflected back into the mixer diode in the correct voltage phase, this power may also be converted to the intermediate frequency, with resultant ideal conversion loss of o dB.

Image-tuned circuits can therefore achieve theoretical low-noise-figure performance, the ultimate performance being restricted only by that of the following i.f. amplifier. However, in practice such circuits are extremely complex and the performance is critically dependent on the circuit conditions. Further, ideal mixer-diode performance is not realised (for image-matched or image-tuned conditions) owing to losses in the diode and at harmonic frequencies.

6.2.4 Conversion loss and noise-temperature ratio

Eqn. 6.2 shows that the conversion loss L_c and the noise-temperature ratio N_r are the mixer-diode parameters determining the overall receiver noise factor. It is also desirable that the diode should have the greatest resistance to electrical overloads compatible with a satisfactory noise factor. A diode rectifier may be represented by the simple equivalent circuit of the barrier resistance R_b in parallel with the barrier capacitance C_b, this combination being in series with the spreading resistance R_s. The actual mixing is accomplished by the nonlinear barrier resistance, the series resistance and barrier capacitance acting as parasitics which degrade the conversion loss. The theory of mixing shows that the conversion loss of the mixer diode is dependent on the low-frequency-rectification properties of the junction, and at any frequency at which the barrier capacitance is significant, the product of the parasitics must be minimised for optimum conversion loss and satisfactory mixer performance.

Thus it may be shown (Torrey and Whitmer, 1948) that

$$L_c \propto C_b R_s \tag{6.3}$$

The contributions to the noise-temperature ratio N_r of the diode are the thermal noise of the series resistance, shot noise of the barrier and flicker noise which is characteristic of the semiconductor junction and surface conditions. At intermediate frequencies in the megahertz range, the major contribution is usually shot noise, but at lower frequencies flicker noise predominates. Noise-temperature ratio is related to the semiconductor surface treatment, the method of forming the junction, the junction area, the ohmic contact and the reverse-current leakage. In practice, a large junction area and a low reverse-current leakage both help to keep the noise-temperature ratio to a minimum.

6.2.5 Resistance to electrical overload (burnout)

R.F. burnout, i.e. when microwave power or spike energy causes degradation of the diode, is a complex function of the device. The r.f. impedance, and hence the absorbed power, the r.f. circuit bandwidth, and the nature of the power (whether c.w., pulsed or t.r.-cell leakage) are involved. Therefore r.f. burnout ratings in terms of maximum c.w. power, peak power or spike energy tend to be rather ambiguous unless the r.f. circuit conditions are carefully defined, and there are inherent difficulties in measuring t.r.-cell spike energy. Although attempts have been made to define diode-burnout ratings by using t.r.-cell spikes simulated by coaxial-line d.c. spikes, these unfortunately have not established a good correlation between r.f. and d.c. performance. Hence actual r.f.-burnout tests give the most reliable information, but it is essential that assessment tests should be carefully defined under specified controlled r.f. circuit conditions to ensure that useful comparative data are obtained.

The physical cause of mixer-diode burnout is also complicated but it is usually accepted that burnout is diffusion or melting caused by high temperatures at the junction.

It can be shown (Torrey and Whitmer, 1948) that

(*a*) for t.r.-cell spike leakage

$$\text{junction temperature} \propto \frac{\text{spike energy}}{(\text{junction radius})^2} \qquad (6.4)$$

(*b*) for pulse power

$$\text{junction temperature} \propto \frac{\text{peak power}}{\text{junction radius}} \qquad (6.5)$$

Microwave-diode burnout is obviously a function of the type of junction, and for a given junction it is dependent on the junction radius which should be as large as is compatible with noise considerations.

6.2.6 Point-contact technology

The conventional point-contact diode and the more recently introduced Schottky-barrier, or so-called hot-carrier diode, are all devices based on the metal–semiconductor junction and only involve majority carriers.

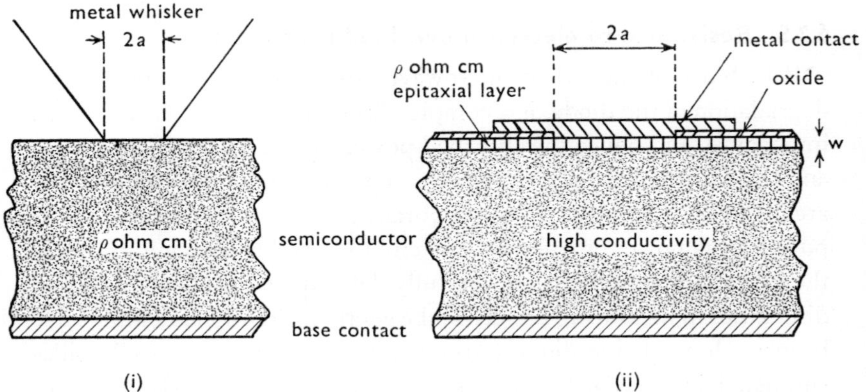

Fig. 27   Point-contact and Schottky-barrier multisemiconductor structures

A model of the conventional form of point contact, assumed to be circular, is shown in Fig. 27(i). The junction area is small compared with the thickness of the semiconductor layer, and the spreading resistance is given by

$$R_s = \frac{\rho}{4a} \tag{6.6}$$

where ρ is the resistivity of the semiconductor and a is the radius of contact, and it may be shown (Messenger and McCoy, 1955) that

$$L_c \propto C_b R_s \propto \frac{a\epsilon^{\frac{1}{2}}}{N^{\frac{1}{2}}\mu} \tag{6.7}$$

where $\qquad \epsilon$ = relative permittivity

$\qquad\qquad N$ = majority-carrier density

and $\qquad\quad \mu$ = carrier mobility in semiconductor material

This expression indicates that the contact radius should be made as small as possible. In practice, this is determined by the design considerations of satisfactory mechanical stability and resistance to electrical overloads.

In general, however, the expression may be taken as a figure of merit for the semiconductor material, but there is usually an optimum value for the product $N^{\frac{1}{2}}\mu$ as the carrier density is increased. So, considering the relative merits of potential semiconductor materials that can be used at normal temperatures, gallium arsenide may be expected to achieve the highest cutoff frequency, followed by germanium and silicon. However, for frequencies up to about X band (for a contact area consistent with point-contact technology) it may be shown that the merits of gallium arsenide and germanium are comparable (Meredith and Warner, 1963).

Point-contact technology is extensively employed to achieve the required small-area junction for microwave-mixer and detector-diode applications. A whisker material has to be chosen which will form a suitable junction and also give the required rectification. In practice, a tungsten-based wire is normally used in conjunction with p type silicon, the junction being mechanically formed. Titanium is normally used with n type germanium and the junction is electroformed. However, the objective is similar in both cases, i.e. to produce an intimate metal–semiconductor contact. Forming the junction has the advantage of control of the barrier capacitance with consequent control of the diode r.f. impedance.

6.2.7 Schottky-barrier technology

The recent development of the Schottky-barrier diode is a result of the rapid advancement in semiconductor planar technology.

The metal–semiconductor junction of the Schottky-barrier device is formed with an evaporated contact rather than a whisker. A model is shown in Fig. 27(ii). This approach maintains the required clean interface between the metal and semiconductor, with resultant low internal-noise generation, and with the further advantage of elimination of possible instability of a whisker pressure contact. The junction area and consequent barrier capacitance are at present larger than can be achieved with point-contact technology; however, by use of epitaxial semiconductor material, the series resistance is reduced to maintain the desired $C_b R_s$ product. For very thin epitaxial layers this product becomes independent of the contact radius as

$$R_s = \frac{w\rho}{\pi a^2} \tag{6.8}$$

where w is the thickness of the epitaxial layer, and it may be shown (Oxley and Summers, 1966) that

$$L_c \propto C_b \propto \frac{w\epsilon^{\frac{1}{2}}}{N^{\frac{1}{2}}\mu} \tag{6.9}$$

Provided, therefore, that an epitaxial layer of the requisite thickness can be achieved and a suitable junction can be formed, the expression suggests further freedom in device design for larger-area—and thus more rugged—devices than with point-contact technology. The use of high-resistivity epitaxial layers will not only reduce the barrier capacitance, but the resultant greater reverse breakdown voltage will increase the dynamic range. An example of a current/voltage characteristic compared with that obtained with point-contact technology is shown in Fig. 28.

The radius of contact of the Schottky-barrier device, compared with about $1 \cdot 0$–$2 \cdot 5 \, \mu$m for point-contact technology, may be about $2 \cdot 5$–$10 \, \mu$m, depending on the semiconductor material, in conjunction with a semiconductor epitaxial-layer thickness of about $1 \, \mu$m.

As already indicated, the Schottky-barrier diode is very similar in concept and in operation to the ideal point-contact diode, and the chief advantages compared with the point-contact device are considered to be: (*a*) improved reliability, i.e. mechanical stability and resistance to electrical overloads, (*b*) I/V characteristics close to those predicted theoretically by Schottky for a metal–semiconductor contact, with consequent nearly theoretical low-frequency conversion loss, (*c*) reduced internal noise generation, with resultant wider dynamic range and (*d*) compatibility with integrated circuits.

A possible disadvantage is the difficulty, in manufacture, of control

of the junction parameters to achieve the necessary reproducibility of
r.f. impedance for use with the pretuned mounts of modern radar
systems.

6.2.8 Backward diodes

The tunnel diode or Esaki diode is a p–n junction diode with
quantum-mechanical tunnelling. The diode itself is made of heavily
doped semiconductor material and is an abrupt junction with an

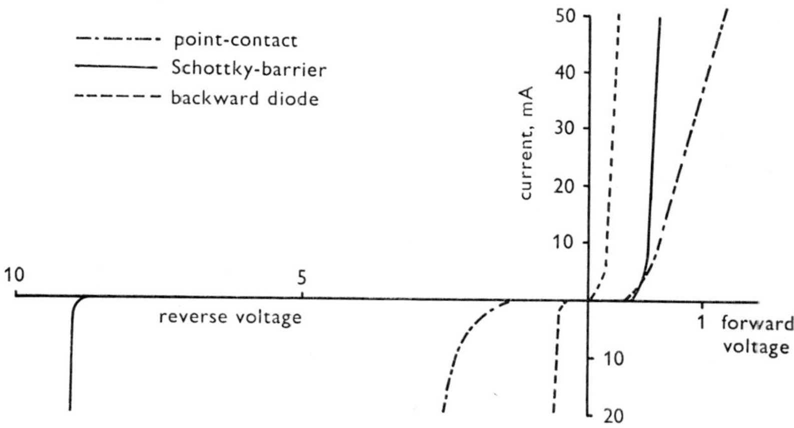

Fig. 28   Example of I/V characteristics of point-contact, Schottky-barrier
and backward-diode microwave diodes

extremely narrow depletion layer. The important characteristic of the
tunnel diode is its negative resistance, but, in the backward diode, the
tunnelling is reduced from the high levels of a conventional diode so
that a negative resistance is almost nonexistent. The tunnel diode,
although capable of microwave mixing and detecting, is difficult to
stabilise and the negative conductance is a circuit problem. A good
compromise is a backward diode which acts as a highly efficient
rectifier over a limited voltage range but in the reverse direction, hence
the term 'backward diode'. This type of characteristic is normally
used for microwave-mixer and detector applications, and Fig. 28

illustrates an example of such a current/voltage characteristic (reversed as used in practice), compared with those that may be obtained with point-contact and Schottky-barrier technologies.

Carrier concentrations of about 10^{19} cm^{-3} are required, and a high mobility is desired to minimise spreading resistance. Germanium, gallium antimonide and gallium arsenide all have acceptable properties. At present, however, microwave backward diodes are normally made from n type germanium with a gallium- or aluminium-doped junction contact.

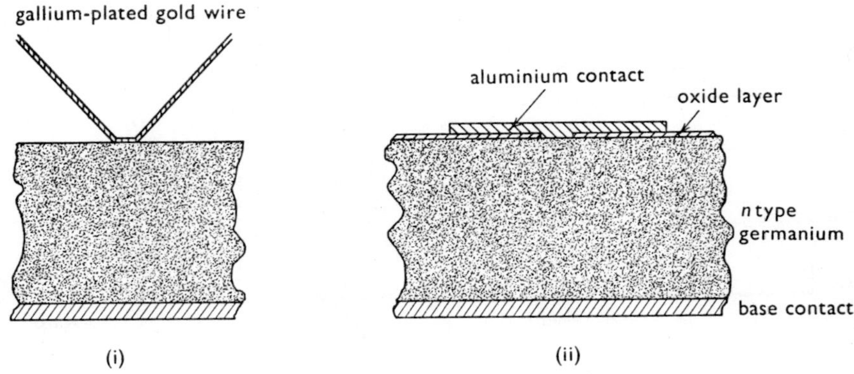

Fig. 29   Pulse-bonded and planar backward-diode structures

Alloyed-junction and pulse-bonded techniques are the chief methods at present employed for the fabrication of microwave backward diodes. The pulse-bonded structure, having small contact area and low capacitance, is attractive for operation at microwave frequencies (Oxley and Hilsden, 1966) and has the advantage of control of the r.f. impedance. However, the larger capacitance of the alloyed junction can be tolerated for microwave application owing to the inherent low series resistance, and the new planar-junction techniques are more compatible with integrated circuits. The construction of such diodes is shown in Fig. 29.

The backward diode offers greater nonlinearity of the I/V characteristic at the origin, good flicker-noise properties and rectifying characteristics. It is relatively insensitive to temperature changes, but has a limited dynamic range (Oxley and Hilsden, 1966). Also, owing to the semiconductor carrier concentration employed, a backward diode may be expected to have a smaller junction area and thus reduced resistance to burnout for the same capacitance, compared with a metal–semiconductor structure.

6.2.9 Overall noise factor for megahertz i.f.

The structure of the basic silicon point-contact diode has remained the same for many years. Optimisation of silicon technology (e.g. out-diffusion by heat-treatment techniques) has produced consistent receiver noise factors of about 12 dB, 8 dB and 7 dB at Q, X and S band frequencies, respectively (including 1·5 dB for an i.f.-amplifier noise factor at 30–60 MHz). By the application of epitaxial techniques, these values have been reduced to about 7 dB and 6 dB at X and S band frequencies, respectively. An alternative approach, using the higher mobility of germanium, has achieved typical receiver performance factors of 8·5 dB, 6·0 dB and 5·5 dB at Q, X and S band frequencies, respectively. By applying epitaxial techniques, it may be feasible to achieve an overall noise factor of about 5·5 dB at X band frequencies. If this figure is realised, the diode performance has approached the theoretical limit, and only by attention to the following amplifier or circuit conditions may further advancement in receiver noise performance be obtained. For example, recent transistor i.f. amplifiers can give noise factors of 1·5 dB at about 45 MHz, and 1·0 dB now seems feasible. Further, image-reflection techniques can give significant improvements in receiver noise factor, although these are not compatible with all systems.

At present, Schottky-barrier diodes attain receiver noise factors of about 5·5 and 6·0 dB at S and X band frequencies, respectively. However, with improved technology, performance figures approaching 5·5 dB at frequencies up to X band will probably be achieved.

Further, because of the nearly ideal I/V characteristics obtained with Schottky-barrier diodes, it is predicted that these figures may be reduced by about 2 dB by image-tuned-circuit techniques.

Bonded-contact germanium backward diodes attain receiver noise factors of about 8 dB at frequencies up to about 18 GHz. Alloyed-junction types attain about 10 dB at X band frequencies. The backward diode, however, is superior to point-contact or Schottky-barrier diodes as a mixer under conditions of limited local-oscillator power, without the complication of bias. Optimum performance is obtained at about 150 μW local-oscillator power, and the diode will operate satisfactorily with powers of less than 30 μW, e.g. in conjunction with tunnel-diode oscillators, but the dynamic range will obviously be limited. For mixer applications with an i.f. in the megahertz range, the backward diode is probably limited to these specialised conditions.

6.2.10 Doppler performance: receiver noise factor for kilohertz i.f.

Although the point-contact diode is a sensitive mixer diode when used with i.f. amplifiers in the megahertz range, it imparts a severe limitation on receiver performance when used with lower-frequency i.f. amplifiers. Although the low-frequency (flicker) noise performance of point-contact diodes has been greatly improved in recent years, from about 35 dB to about 20 dB at 10 kHz and X band frequencies, respectively, the performance is still degraded by about 15 dB compared with the performance at megahertz frequencies. It appears unlikely that further significant improvements in this respect will be made with these diodes.

Information on the low-frequency noise performance of the Schottky-barrier diode is rather limited at this early stage of its entry into the mixer-diode field. However, initial performance evaluation has indicated that it is better than the conventional point-contact device and almost as good as the backward diode, and therefore may prove to be very suitable for this application.

The backward diode, whether of the bonded-contact or alloyed-junction type, has a low flicker noise. X band-receiver noise factors in the range of 10–15 dB may be obtained at an intermediate frequency of 10 kHz, a degradation of less than 5 dB compared with the performance at megahertz intermediate frequencies. Unfortunately, however, the narrow dynamic range has tended to limit the use of backward diodes in practical systems and it is likely that the Schottky-barrier diode will prove superior for this application.

6.2.11 Burnout

The principal requirements for resistance to burnout of mixer diodes is the ability to withstand spike leakage from a t.r. cell. The majority of t.r. cells are designed to give a breakthrough of about 0·2 erg/spike at X band frequencies and 0·05 erg/spike at Q band frequencies (assuming an r.f. bandwidth of about 10 %), and although there are problems of unreliability and unexplained diode burnout, the majority of point-contact mixer-diode types are normally considered to have a satisfactory life under controlled conditions for these levels. However, development of solid-state limiters should give greatly increased receiver reliability, and these t.r.-cell–varactor limiters can already reduce the spike leakage to about 0·01 ergs at X band frequencies.

Little is known about the burnout capability of the Schottky-barrier diode, for t.r.-cell leakage. Care must be taken when using comparative data, for single d.c. spike or multiple d.c. spike tests may be used for burnout evaluation, and these may not necessarily correlate with r.f.-burnout performance. The larger contact area of the Schottky-barrier diode compared with that of the point-contact diode would certainly suggest improved burnout capability and possibilities for improved receiver reliability.

The burnout performance of the backward diode appears to be independent of the junction structure, i.e. bonded contact or alloyed, but, as may be expected, it is markedly dependent on the junction area. Diodes developed for Doppler applications are considered to have burnout performance comparable with that of the conventional point-contact device, but reduced-capacitance diodes developed specifically as broadband detectors require t.r.-cell–varactor limiters to afford adequate protection for mixer applications. However, special types could be developed with improved burnout capability, i.e. more than 0·4 erg/t.r. spike coupled with receiver noise factors of about 9 dB, but these would have such inherent disadvantages as a narrow dynamic range and a low i.f. impedance of about 50 Ω.

6.2.12 R.F. impedance and bandwidth

The r.f. impedance and the bandwidth of mixer diodes are functions of the junction parameters and the encapsulation parasitics.

Types of encapsulation in current use are shown in Plate 16.

The whisker inductance of the point-contact structure plays a significant part in diode design and may be used to advantage for r.f.-impedance adjustment. The X band cartridge of the 1N23 type is designed primarily for use across waveguide, and its r.f. impedance is fairly high (about 400 Ω), whereas the reduced inductance associated with the recently introduced miniature double-ended cases results in a relatively low r.f. impedance (about 50 Ω at the design frequency), with consequent versatility for waveguide, coaxial, stripline applications and wide frequency range. Coaxial configurations are normally designed for compatibility with about 70 Ω r.f. impedance. The r.f. impedance of the backward and Schottky-barrier diodes in a low parasitic encapsulation is likely to be lower; about 5–10 Ω for the backward diode, and 50 Ω for the Schottky-barrier diode. However, as already indicated, this is critically dependent on the encapsulation parasitics.

The normal system r.f. bandwidth requirement of about 10% is usually met satisfactorily with point-contact devices. However, with the introduction of miniature-case types and their associated reduced whisker inductance, the potential bandwidth is increased to about 30% or better. Although the Schottky-barrier or alloyed backward diode may be encapsulated in a low-inductance case, the expected bandwidth advantage may not be obtained owing to the increased junction capacitance likely to be associated with these types. However, of these, the Schottky-barrier type has the best possibilities for improvement and subsequently reduced capacitance.

6.2.13 I.F. impedance

The i.f. impedance of the conventional point-contact diode is normally about 300–500 Ω and is fairly dependent on local-oscillator drive. The Schottky-barrier diode offers a lower impedance, about 100–400 Ω, which is also fairly dependent on local-oscillator drive. The backward diode has the lowest value, about 50–150 Ω, but, in this case, the dependence on local-oscillator drive is greatly reduced.

6.2.14 Cooled mixer diodes

It is not common for mixer diodes to be operated at reduced temperatures, but Alfeyev (1966) has described measurements which suggest that further noise improvements are possible if the diode is cooled. With a silicon diode, the measured noise performance improved by 0·5–2·0 dB on cooling the diode to 200 °K, but further cooling produced a rapid deterioration in performance.

On the other hand, the noise performance of mixers using GaAs, and InAs diodes continued to improve on cooling to 77 °K. Below about 150 °K it is possible to use InSb also, and, in this temperature range, the lowest measured noise factors were obtained for diodes of this material. The noise factor of the InSb mixer at 77 °K, measured in the 8–10 cm band was 4 dB lower than that for a conventional uncooled silicon diode.

For obvious practical reasons, the use of cooled mixers is likely to be restricted, but there may be some applications where the mixer can conveniently be incorporated with a cooled parametric amplifier.

6.3 Detector diodes

A noise factor cannot be defined (as for a superheterodyne receiver) for a crystal-video receiver. Instead, a signal level is found such that the output-signal power is just equal to, or at a specified level above, the output-noise power.

6.3.1 Crystal-video receiver performance

The quality of the crystal-video detector can be expressed quantitatively in terms of short-circuit current sensitivity and video impedance, by the 'figure of merit', which may be defined as

$$M = \beta Z_V^{\frac{1}{2}} \tag{6.10}$$

where $\qquad \beta$ = microwave current sensitivity

and $\qquad Z_V$ = video impedance

However, the expression does not include the noise properties of the detector and does not present the true quality of devices under conditions of forward bias.

A widely used performance criterion for video detectors is 'tangential sensitivity', which indicates the ability of the detector to detect a signal against a noise background, and includes the noise properties of the detector and video amplifier.

The tangential sensitivity performance, in watts, may be derived from the following equation (Lucas, 1966, and Frohmaier, 1960):

$$S_t = \frac{1}{\beta\sqrt{Z_V}}\{4\,YkTB(F_v+t_v-1)\}^{\frac{1}{2}} \qquad (6.11)$$

where

k = Boltzmann's constant

T = temperature, °K

B = video-amplifier bandwidth, Hz

F_v = video-amplifier noise factor

t_v = noise-temperature ratio of detector diode

and Y = 6·3, an experimentally determined factor, relating the edge-to-edge height of a band of Gaussian noise to its r.m.s. value

6.3.2 Diode sensitivity

The diode microwave current sensitivity may be given (Torrey and Whitmer, 1948) by the relation

$$\beta = \frac{\beta_0}{1+\omega^2 C_b^2 R_b R_s} \qquad (6.12)$$

where β_0 is the low-frequency current sensitivity determined by the curvature of the I/V characteristic at the operating point.

Thus, similar considerations apply to detector and mixer diodes, namely that, for high efficiency, C_b and R_s must be minimised. How-

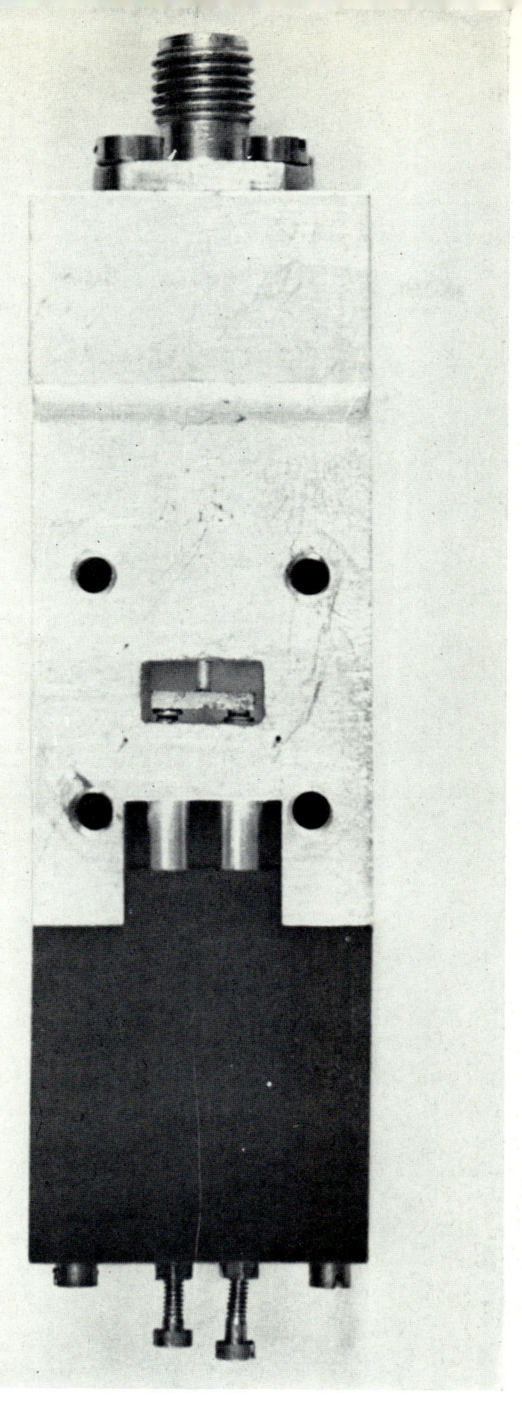

11   Parametric amplifier using two micropill diodes separated by
a length of microstrip line

(Photo: *Philips Technical Review*)

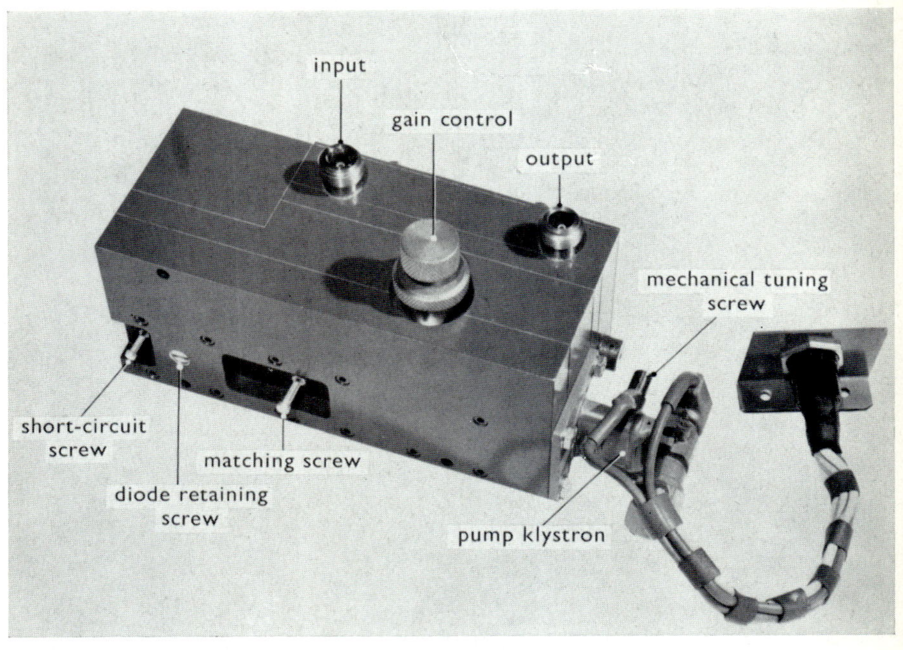

12 Compact S band parametric amplifier using
balanced-diode configuration.

(Photo: Ferranti Ltd.)

13   Model showing construction of varactor diode

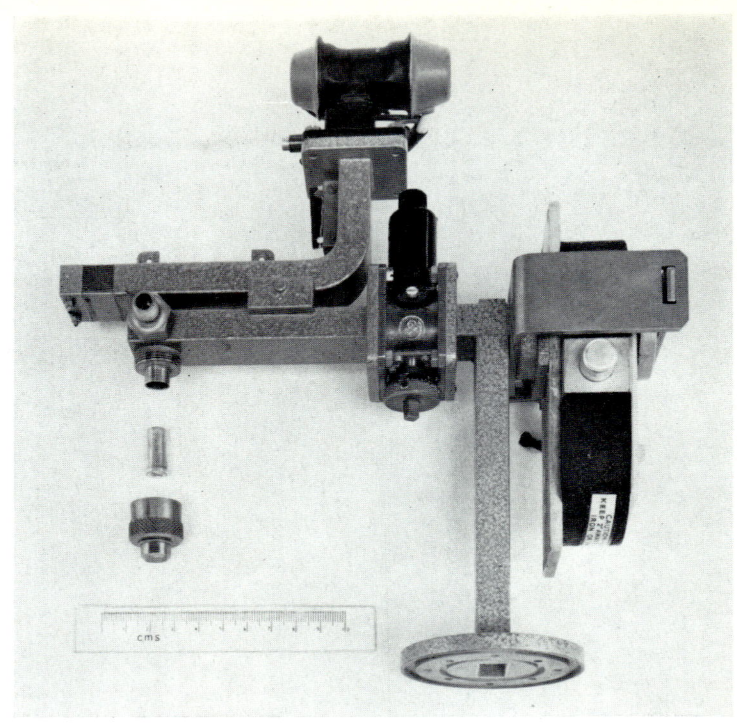

14   Simple X band transmit–receive system
(i.f. amplifier not included)

(Photo: Kelvin Hughes Ltd.)

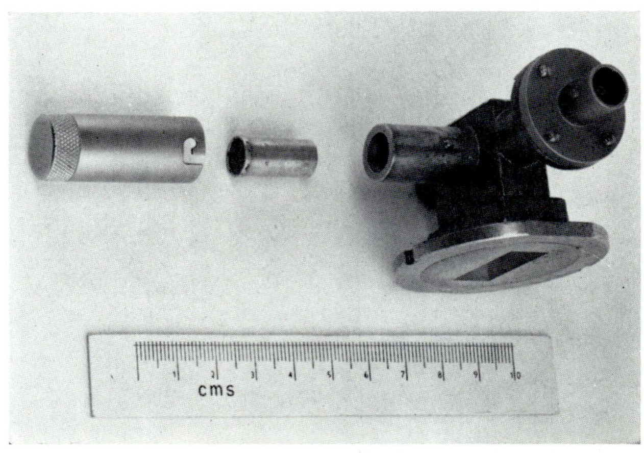

15   Simple X band crystal-video receiver
(video amplifier not shown)

16 Examples of microwave-mixer and detector-diode encapsulations

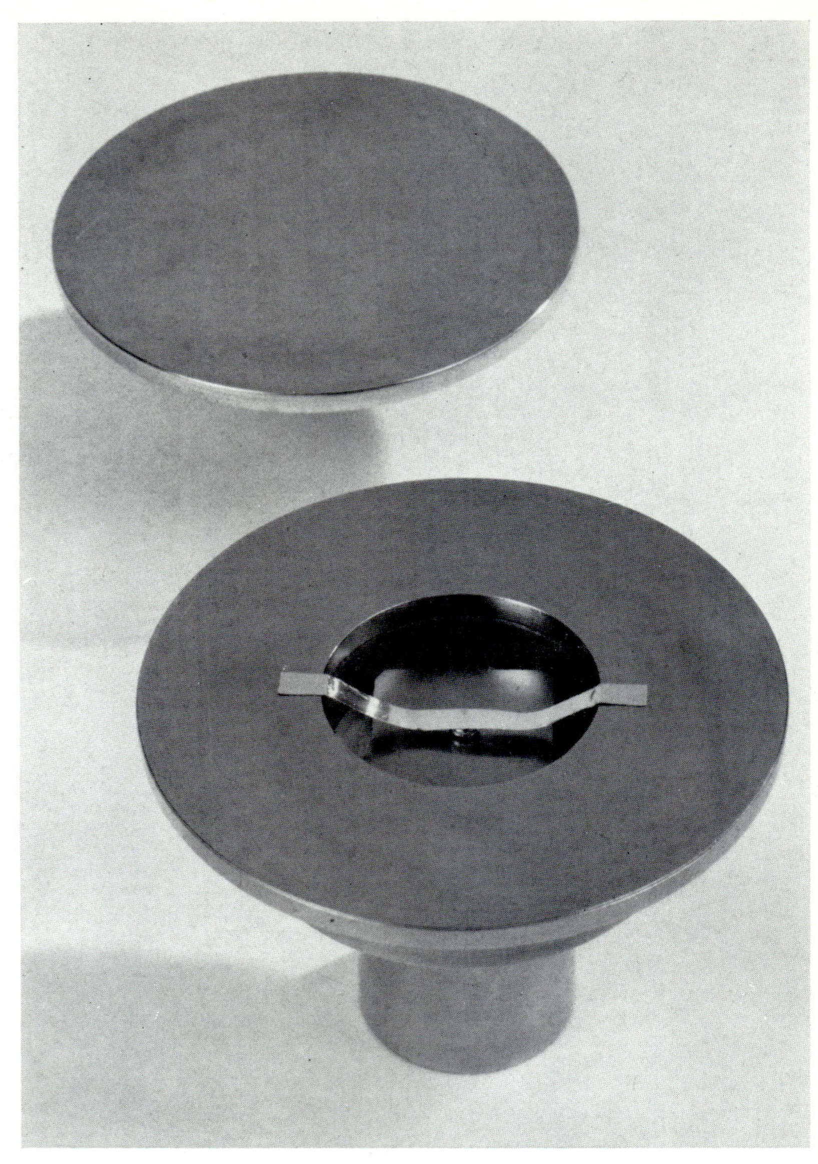

17 Construction of typical packaged tunnel diode

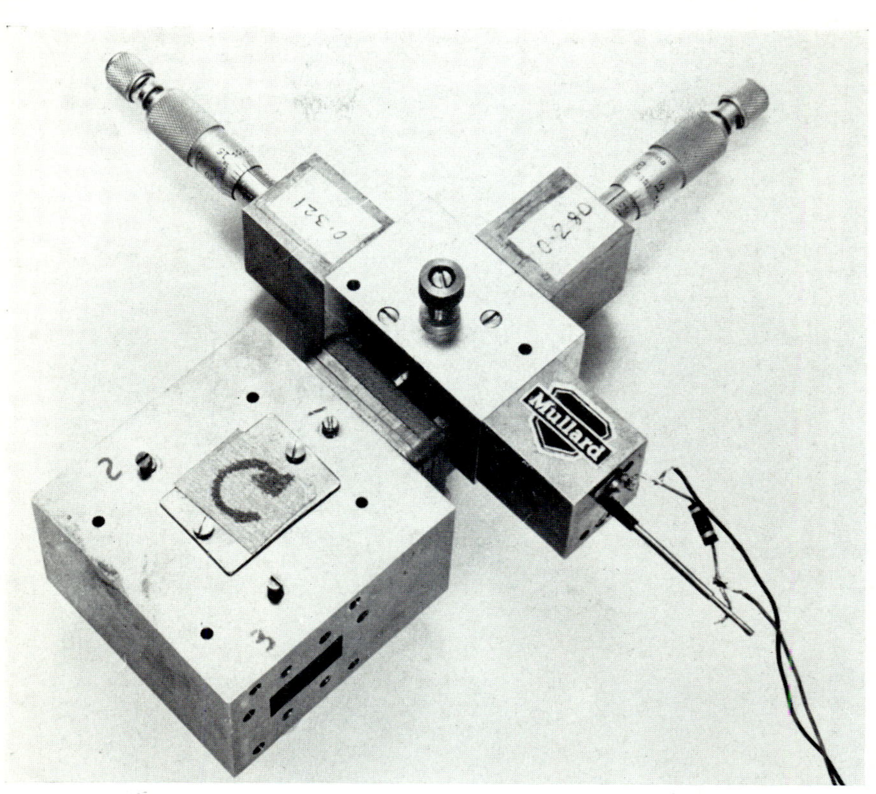

18   Complete tunnel-diode amplifier

19   Stripline and l.i.d. microwave-transistor packages

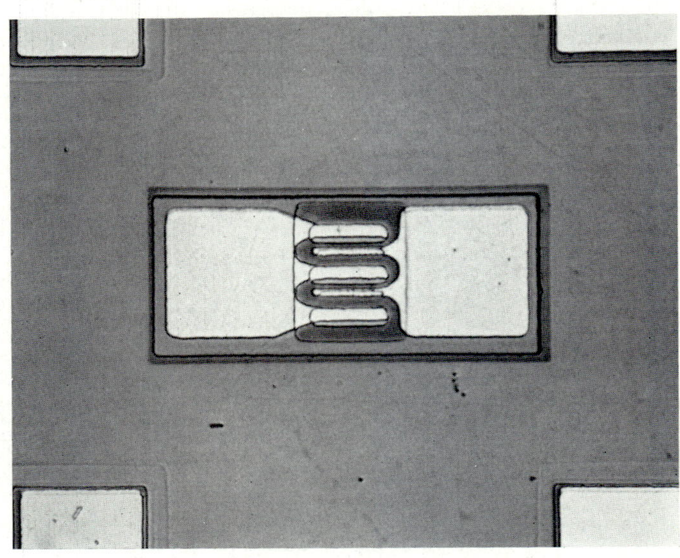

20   Low-noise silicon-planar microwave transistors
with 1μm-wide emitter finger ($\times 700$)

ever, detector sensitivity depends on the square of C_b in contrast to linear dependence for the mixer.

For detector applications, the best type of I/V characteristic is one with maximum curvature at the operating point; it is desirable that this should occur at zero bias, as the use of bias current may introduce some additional noise into the systems.

Owing to the greater curvature at the origin and the steeper I/V characteristic associated with the backward diode, the backward diode may be expected to have a potentially higher current sensitivity than that of a point-contact diode, and, as the curvature occurs at the origin, optimum current sensitivity may be expected at zero bias.

The best conventional point-contact Ge and Si detectors, biased for optimum sensitivity, achieve tangential sensitivities, for a 1 MHz video bandwidth, of about -52 dBm at X band frequencies and -47 dBm at Q band frequencies. Comparable video-impedance backward diodes can offer an improvement of about 5 dB at X band frequencies and about 8 dB at Q band frequencies, and this may be obtained without the complication of bias. Sensitivities better than -60 dBm may be achieved at X band frequencies with a video impedance of about 5 kΩ, using bonded-backward-diode techniques, and zero bias operation offers significant advantages in low-frequency applications where flicker noise predominates.

Schottky-barrier diodes are unlikely to give better sensitivities than those of backward diodes, and bias will be required to optimise performance. X band sensitivities of about -52 dBm are achieved at present.

6.3.3 R.F. bandwidth

The bandwidth of a detector diode is of the order of 40 % and is determined by the compromise between sensitivity and resistance to burnout required to achieve the best overall performance. It is reasonable to expect better bandwidths for low values of barrier capacitance, and for low values of video impedance, as the relative effect of the barrier capacitance is reduced. Conventional detector diodes are, in fact, often biased in the forward direction, not only to take advantage of any improvements in sensitivity but also to reduce their video impedance, with a consequent improvement in bandwidth.

The backward diode exhibits low video impedance and a small junction capacitance may be achieved with bonded-contact structures, and so this device offers the best potential broadband properties. A backward diode in a miniature capsule gives a bandwidth of 2–18 GHz with a tangential sensitivity of -54 dBm over this frequency range.

6.3.4 Resistance to electrical overloads (burnout)

Similar considerations apply for detector and mixer diodes in this respect. However, overloads for detector applications consist of microsecond pulses, and it is the peak pulse power rather than the pulse energy considered for mixer diodes that determines the extent of burnout.

It is desirable to maximise the junction (or contact) radius consistent with satisfactory r.f. performance. The junction (or contact) temperature may be shown to be linearly dependent on the radius for a given junction in contrast to the square of the radius for mixer spike burnout.

The X band burnout level of backward diodes may cover the range 1–20 W* compared with 1–3 W peak power for point-contact devices, depending on the application; for instance, a low burnout level is associated with high sensitivity and broadband operation, and a high burnout level is associated with medium sensitivity and narrowband operation. There is no information available yet for Schottky-barrier devices in this respect. However, the larger contact area of this device should give it some advantage over the point-contact diode.

6.4 Hot-carrier microwave detector

A new microwave detector has been introduced based on the thermo-electric effect of hot carriers in bulk semiconductors (Harrison and Zucker, 1966).

Semiconductors subjected to sufficiently high microwave-power-density gradients exhibit an open-circuit voltage which arises from the

* Incident peak power under pulsed conditions. The absorbed power will be of the order of 10 mW

nonuniform heating. This gives rise to an e.m.f. of which the average component is proportional to the applied r.f. power. The nonuniform applied electric field and power density are established by a point-contact high–low junction. In such a contact, the semiconductor at the point of contact is of the same conductivity type but more heavily doped than the bulk material. This may be achieved by doping either the metal point or the semiconductor surface.

Experimental germanium detectors based on this principle and produced for the 50–100 GHz frequency range have sensitivities comparable to that of the conventional point-contact device, but with greatly improved burnout and stability qualities.

6.5 Microwave-mixer and detector diodes: summary and future trends

Conventional point-contact mixer diodes at present available are capable of providing low noise factors for intermediate frequencies in the megahertz range, and, with current developments, are likely to approach their theoretical limitations at frequencies up to about X band. However, as the overall receiver performance now depends to a significant extent on the noise factor of the following i.f. amplifier, further attention here could benefit receiver performance. Improvements in noise factors of point-contact diodes may be associated with reduced resistance to electrical overloads, and so good protection devices are necessary. This requirement is being met by development of t.r.-cell–varactor-limiter combinations. Future development of point-contact technology is therefore likely to be restricted to frequencies above X band, where there is still scope for improvements in diode performance.

The Schottky-barrier diode is the first diode made by techniques that do not require a wire in direct contact with a semiconductor surface and is likely to replace existing types. Although it is unlikely to improve receiver noise-factor performance for intermediate frequencies in the megahertz range, the Schottky-barrier diode offers several potential advantages over point-contact types, chiefly, improved stability, improved resistance to electrical overloads, wide dynamic range and compatibility with integrated circuits. Initial results indicate that it also has good flicker-noise properties, and, coupled with the dynamic range, this suggests that the device has a high potential for Doppler applications. Improvements in the fabrication technology can be envisaged which will increase the operating frequency of this type of diode.

The microwave applications of the backward diode are as a mixer with moderate noise factor in conjunction with limited-power-output local oscillators, as a Doppler mixer and as a video detector. Although the flicker-noise properties of this type are superior to those of the point-contact and Schottky-barrier diodes, poor dynamic-range performance tends to limit its use in practical Doppler systems.

The main potential of the backward diode is as a low-level detector. The conventional point-contact rectifier has long played an important role for video-receiver applications, and although it is still the most widely used form of detector, backward diodes are proving competitive. In general, the advantages of the backward-diode detector are due to the low video impedance (allowing wide r.f. and video bandwidths) and relatively high sensitivity coupled with an r.f. impedance compatible with 50 Ω systems. A significant advantage is also obtained in low-frequency applications where flicker noise predominates, owing to the high sensitivity obtained without bias. The application of planar techniques could produce rugged devices compatible with integrated circuits.

The hot-carrier diode based on the generation of a thermoelectric voltage by high microwave-power densities may have limited possibilities for rugged low-level detector applications at millimetre wavelengths.

7 TUNNEL-DIODE AMPLIFIERS

7.1 Introduction

In 1957, Esaki (1958) discovered that quantum-mechanical tunnelling in a degenerate-semiconductor p–n junction could provide a negative-resistance characteristic. The tunnel diode was then rapidly developed, primarily for switching applications, and soon reached a high degree of sophistication. However, it has not been adopted for the new generation of computers and in practice the major use is in relatively low-noise microwave amplifiers. In this application it has considerable advantages, being of a very simple construction and offering reasonable noise factors of 3–6 dB combined with good bandwidth. The early limitations of the tunnel-diode amplifier (t.d.a.) were its low power-handling capability and a tendency to fragility unless a good device design was achieved, but both these aspects have improved steadily during the past few years.

The tunnel diode has the voltage-controlled negative-resistance characteristic shown in Fig. 30, in which the main parameter is the peak current. The most useful property of the diode is that the negative-resistance part of the characteristic results from quantum-mechanical tunnelling, and is independent of frequency throughout the microwave spectrum.

In the tunnelling process, electrons can penetrate the high-electric-field region associated with the p–n junction barrier. This can only take place in degenerate or very heavily doped semiconductors in which the bottom of the conduction band on the n type side of the junction is lower than the top of the valence band on the p type side. To increase the tunnelling probability it is also necessary to have a sharp discontinuity between the n and p type materials in order to reduce the width of the transition region to 50–100 Å. With a slight forward bias applied to such a diode, there is an overlap between the relevant conduction and valence bands and a large tunnelling current flows. As the bias is increased, there is less overlap and the current decreases until the normal forward characteristic of a p–n junction is resumed.

When suitably biased in the negative-resistance region, the tunnel diode has the equivalent circuit of Fig. 31 a (Hall, 1960). The junction of the tunnel diode is represented by a negative resistance $-R_n$ (or

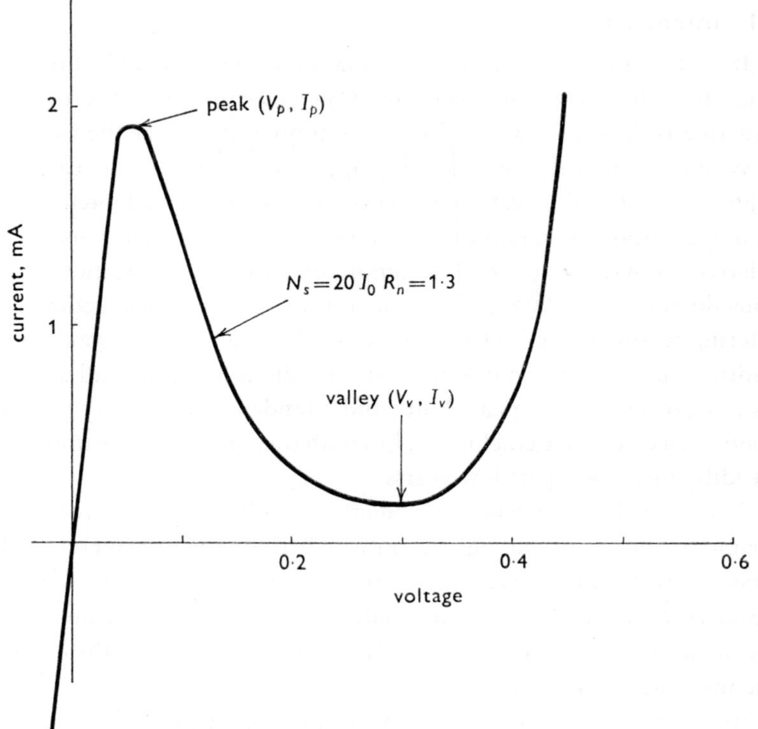

Fig. 30 Typical voltage/current characteristics of germanium microwave tunnel diode (Mullard type AEY 15)

conductance $-G_n$) in parallel with the junction capacitance C_j, where R_n is a positive number. R_s and L represent the unavoidable resistance and inductance in series with the diode.

The diode impedance $Z(\omega)$ in Fig. 31 a is given by

$$Z(\omega) = R_s - \frac{G_n}{G_n^2 + \omega^2 C_j^2} + j\left(\omega L - \frac{C_j}{G_n^2 + \omega^2 C_j^2}\right) \qquad (7.1)$$

The diode can be used as a negative-resistance amplifier at any frequency below the resistive cutoff frequency f_{r0}, at which point $\mathrm{Re}\{Z(\omega)\} = 0$. The series, or self-resonant, frequency f_{x0} gives $\mathrm{Im}\{Z(\omega)\} = 0$. Provided that

$$L < C_j R_n R_s \tag{7.2}$$

$$f_{x0} > f_{r0} \tag{7.3}$$

and it will be relatively easy to design a stable amplifier using the diode. Under these conditions, the simple equivalent circuit of Fig. 31b is a close approximation for frequencies from zero to $f_{r0}/3$, and R and

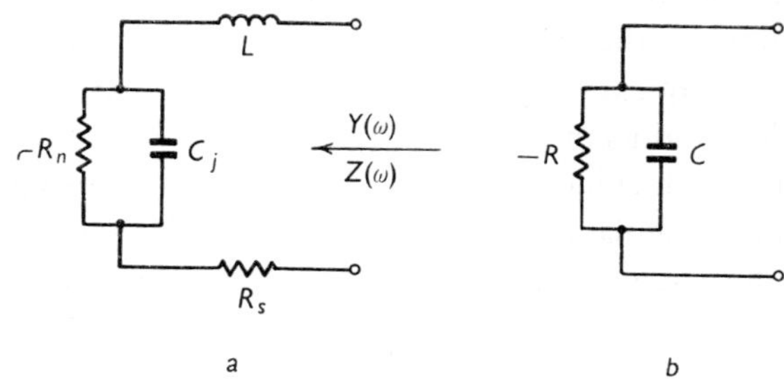

Fig. 31   Equivalent circuits of tunnel diode

a   Equivalent circuit
b   Approximate equivalent circuit, valid for $f < f_{r0}/3$

C differ little from R_n and C_j (McPhun, 1967). This simple equivalent circuit of constant negative resistance in parallel with constant capacitance can be used as a sound basis for initial amplifier design, but the full equivalent circuit must be used when considering out-of-band performance, which affects amplifier stability (Scanlan, 1966).

7.2 Construction of tunnel diodes

At present most diodes are fabricated by alloyed-junction techniques, in which the junction is formed by alloying a small n-doped bead into the surface of a p-type semiconductor wafer. The wafer is soldered into the housing and the connection soldered to the bead. A bead of about 25 μm diameter is used in order to keep the junction capacitance low, the capacitance being directly proportional to the wetted area of the alloyed bead. The junction capacitance is further reduced by electrolytic etching of the junction, which also allows the peak current to be reduced to the desired value.

Etching to achieve much less than 1/20 of the initial peak-point current clearly gives a weak structure, and it is apparent that diodes having resistances above a few ohms are likely to be quite fragile. These considerations lead to the choice of a low etching ratio (ratio of initial to final peak-point current) and this results in a satisfactory mechanical and environmental life-test performance.

The housing in which the tunnel diode is mounted is of considerable importance since it contributes the major part of the series inductance and capacitance of the diode. A low series inductance results in a high self-resonant frequency f_{x0} for the diode, and cases have been designed to give a low inductance by using a thin, small-diameter, ceramic annulus. The inductance can be further reduced by using short connections of rectangular cross-section to connect the p–n junction to the housing. The use of rectangular-cross-section foil can also increase the mechanical strength of the device in its weakest direction, at right angles to the foil in the plane of the wafer surface. Other methods of reinforcing the junction by using glass encapsulation and similar techniques are also employed. These designs have resulted in series inductances of approximately 100 pH, and case capacitances of 0·25 pF. The construction of such a diode is shown in Fig. 32 a and Plate 17.

Silicon, germanium, indium-antimonide and gallium-antimonide tunnel diodes have all been fabricated by this basic alloyed-junction method, and a wide range of devices are available. For low-noise amplifiers, gallium-antimonide diodes can ideally offer a better noise performance, but, since gallium-antimonide technology is less well advanced, most diodes are based on germanium.

It would appear that diodes based on alloying techniques are now virtually fully exploited and new technologies must be used to obtain improved performance. The fragility of the etched-mesa junctions produced in the present alloyed diodes restricts their use to X band frequencies and lower. Because the encapsulation forms an essential part of the strength of the device, these diodes are not suitable for

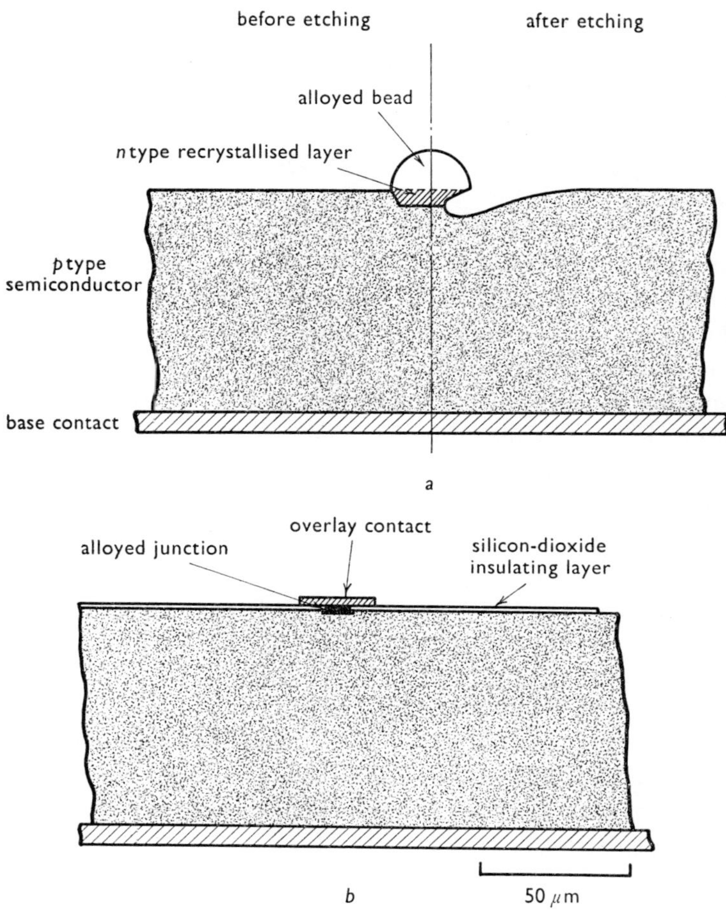

Fig. 32　Tunnel-diode construction

　　　　a　Alloyed device

　　　　b　Planar device

incorporation into an integrated-circuit type of amplifier, and so the true junction performance, which is degraded by the parasitics of the encapsulation, cannot be realised. However, planar technology has advanced rapidly in the past few years, and it is now possible to consider a planar tunnel diode in which the junction is defined by an oxide mask without subsequent etching or processing. It appears likely that germanium will continue to be the most suitable material, and oxide masking and photolithography can now be used to fabricate junction areas of a few micrometres diameter. Diodes have been made by this process with f_{r0} up to 50 GHz, and, although the yield is low at present, this is an important change in technology. The structure of this type of diode is shown in Fig. 32 b.

Devices of this type are significant in offering improved mechanical strength compared with existing mesa diodes. As planar methods improve, and possibly with the introduction of electron-beam techniques, it is becoming possible to produce smaller junctions leading to satisfactory amplifiers for frequencies above X band. These diodes are in chip form, free of encapsulation, and can be incorporated into thin-film hybrid circuits. Although a completely monolithic microwave integrated circuit cannot be formed in germanium, it is now possible to grow germanium islands in a semi-insulating gallium-arsenide substrate by means of contour epitaxy. With this isolation, and deposited transmission lines, a monolithic tunnel-diode amplifier circuit becomes feasible.

7.3   Amplifier design

The use of the circulator-coupled reflection amplifier is now universal for the t.d.a., as this has a wider bandwidth and a lower noise factor and is less dependent on load and generator impedances than a t.d.a. without a circulator.

The basic form of a circulator-coupled t.d.a. is as shown in Fig. 33 a, where the circulator isolates the reflection amplifier from both the generator and the load. The reflection amplier shown in Fig. 33 b terminates the transmission line of characteristic resistance R_0 in a negative resistance $-R$, giving a reflection coefficient Γ which is greater than unity:

$$\Gamma = \frac{R+R_0}{R-R_0} \qquad (7.4)$$

The power gain of the circulator-coupled amplifier is then Γ^2.

To use a tunnel diode as a reflection amplifier, the designer must do the following four things:

(*a*) bias the diode, using a load line providing a single-valued inter-
section in the negative-resistance region of the I/V characteristic

(*b*) present the transmission line with a suitable negative resistance
over the passband

(*c*) reduce Γ to acceptable limits outside the passband

(*d*) maintain stability.

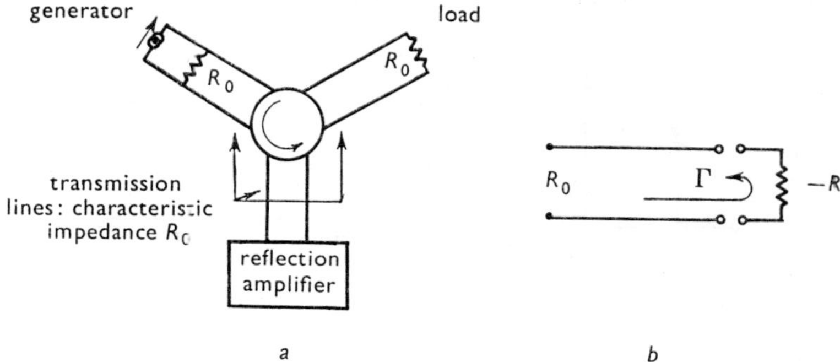

Fig. 33 Details of tunnel-diode amplifier
   a Basic circulator-coupled amplifier
   b Reflection amplifier

Because the tunnel diode presents a negative resistance at its
terminals from d.c. to the resistive cutoff frequency, oscillations may
occur anywhere within this range, so the maintenance of stability has
been a major problem for the circuit designer. Oscillations may occur
owing to resonance within the reflection amplifier, or owing to interactions with the circulator, generator and load. Eqn. 7.4 will only
apply to a 'perfect' circulator with characteristic resistance R_0 at all
frequencies. All practical circulators have a limited passband, and
oscillations are most likely at the band edges.

To avoid generator and load interactions, 4- or 5-port circulators are
frequently used. It is now common practice to provide a stabilising
circuit either in parallel or in series with the diode. The stabilising
circuit provides a resistive load for the diode at all frequencies except
for the desired passbands. However, by using a sufficiently wideband
circulator, it is just possible to design a satisfactory amplifier without
a stabilising circuit. A basic lumped-element stabilising circuit of the

parallel type is shown in Fig. 34*a* and a transmission-line version in Fig. 34*b*. These have the general form of a bandstop filter interposed between the diode and the stabilising resistance R_{stab}.

A matching network is often required between the circulator and diode to give the required bandpass characteristics, and the complete t.d.a. has the form shown in Fig. 35. It is often convenient to include the bias circuit in with the stabilising circuit.

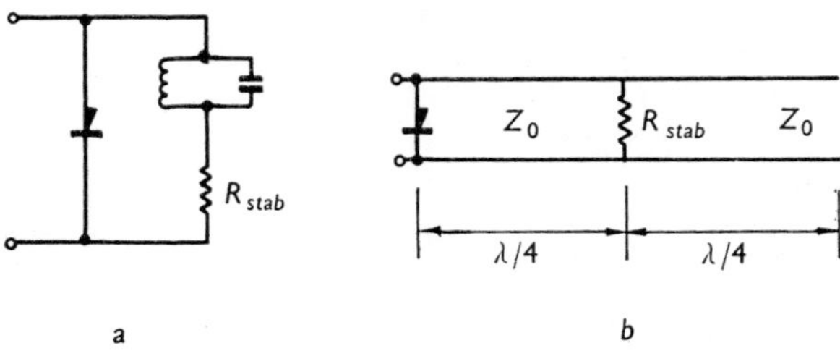

Fig. 34   Shunt-type stabilising circuits
 a Lumped element
 b Transmission line

It is evident that all the frequency-variable circuit elements will together determine the amplifier frequency response. However, much useful theory and design has been based on the assumption of the 'perfect' circulator (Scanlan and Lim, 1964 and 1965, and Hamasaki, 1965). This assumption is not now required, as Okean (1966*a*) has described a powerful synthesis technique which takes into account all the circuit elements including the bias circuit and the circulator.

T.D.A.s have been constructed in waveguide, coaxial and stripline, and there are many papers giving constructional details, some examples being Hamasaki (1965), McPhun (1966 and 1967), Lee (1967), Easter (1965) and Okean (1966*b*). A typical amplifier is shown in Plate 18.

7.4   Amplifier performance

Here we will review the factors affecting attainable t.d.a. performance, and examine what has been achieved so far.

124

7.4.1 Gain

The gain may be set to any value up to infinity by a suitable choice of
the circulator impedance and the negative resistance of the reflection
amplifier. However, amplifiers with gains in excess of 20 dB are likely
to be unduly sensitive to changes in circulator impedance with temperature. Also, saturation will occur at a lower input-signal level for a

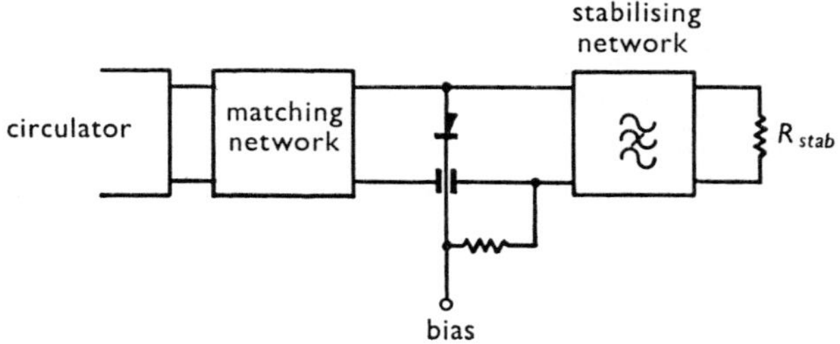

Fig. 35   General form of tunnel-diode amplifier

high-gain amplifier. When a t.d.a. is used as a preamplifier, the gain
will be chosen just sufficiently high to render the noise contribution
from the second stage negligible. A gain of 15 dB is typical.

7.4.2 Bandwidth

The tunnel diode is an inherently wideband device. Assuming the
simplified equivalent circuit of Fig. 31 b, the gain–bandwidth product
for a simple tuning circuit is

$$G^{\frac{1}{2}}B = 1/(\pi RC) \qquad (7.5)$$

and this may be improved by the addition of more reactive elements.
The theoretical limit for an infinite number of tuning elements yields
the bandwidth

$$B = \frac{4\cdot34}{RC(10\log_{10}G)} \qquad (7.6)$$

and this limit is closely approached by six elements.

In practice, the bandwidth is also limited by the diode series
resistance and package capacitance C_s. Lead inductance imposes a

further limit on the range of gain (minimum and maximum) realisable for a given bandwidth, and vice versa (Scanlan and Lim, 1965, and Okean, 1966).

Typical figures for an X band germanium tunnel diode are $C_j = 0\cdot18$ pF, $R_n = 77\ \Omega$, $C_s = 0\cdot25$ pF and resistive cutoff frequency $f_{r0} = 40$ GHz. For an amplifier used at frequencies well below f_{r0}, an effective junction capacitance of $C_j + C_s$ can be assumed, giving the predicted figures shown in Table 4.

The reactance of the circulator may be used to form one or more of the above tuning elements.

Amplifiers with 1 GHz bandwidth at 20 dB gain are available from a number of manufacturers.

Table 4

Gain dB	Bandwidth with simple tuned circuit GHz	Maximum theoretical bandwidth GHz
10	3	13
20	1	6·5

7.4.3 Noise factor

The tunnel diode exhibits shot noise at the junction and also a small amount of thermal noise owing to the bulk series resistance R_s. The latter noise becomes significant only for the higher-frequency amplifiers at X band frequencies and above, where the operating frequency may be in the region of $f_{r0}/3$ or higher.

Nielson (1960) shows the noise factor F to be

$$F = \frac{1 + N_s}{\left(1 - \dfrac{R_s}{R_n}\right)\left\{1 - \left(\dfrac{f}{f_{r0}}\right)^2\right\}} \tag{7.7}$$

It is therefore mainly dependent on the noise constant N_s, which is related to the negative characteristic as

$$N_s = \frac{e}{2kT} I_0 R_n \simeq 20 I_0 R_n \quad \text{(at 290 °K)} \tag{7.8}$$

where
$$I_0 = \tfrac{1}{2} I_P$$

As the peak current I_P is inversely related to R_n, this is a function of the basic properties of the semiconductor material (Armstrong, 1962), and

$$I_0 \propto \exp A\left\{E_g\left(\frac{\epsilon m}{N}\right)^{\frac{1}{2}}\right\} \tag{7.9}$$

Therefore an appropriate material has a small band gap E_g and a small effective mass m of the lighter carrier, to give a high tunnelling probability. It should also possess a low relative permittivity ϵ, and be capable of being doped to a high level, as N is the weighted average carrier concentration on the two sides of the junction. A high carrier mobility is also desirable, to reduce the series resistance R_s. Table 5 shows the basic material parameters of some of the more promising semiconductors compared with the achieved noise factors and theoretical values, assuming that $f \ll f_{r0}$ and $R_s \ll R_n$.

Table 5

Material	ϵ	E_g eV	m	Electron mobility μ cm²/Vs	N_s	Theoretical noise factor dB	Achieved noise factor dB
Ge	16·0	0·68	0·55	3000	1·28	3·5	$\begin{cases}4 \text{ at 2 GHz} \\ 5\cdot5 \text{ at 9 GHz}\end{cases}$
Si	11·8	1·08	1·1	1000	—	—	—
GaAs	12·5	1·40	0·034	5000	1·4	5·5	—
GaSb	14·0	0·70	0·047	4000	0·85	2·5	$\begin{cases}3\cdot2 \text{ at 2 GHz} \\ 4\cdot7 \text{ at 9 GHz}\end{cases}$
InSb	17·0	0·18	0·021	100000	0·5*	2	—

* Below about 100 °K

The minimum value of N_s quoted above occurs for a bias voltage somewhat greater than that at the inflection point of the I/V characteristic. Thus, for minimum F, the $R_n C_j$ product is not a minimum, so that bandwidth must be sacrificed if the minimum noise factor is to be obtained. Bandwidth may decrease by 30 % as bias is increased from the inflection point to the point of the minimum noise factor.

The intensive study of tunnel-diode design and fabrication techniques, mainly for computer applications, has benefited the devices for t.d.a.s, and these have virtually achieved the theoretical noise factors.

InSb devices can only be used below 77 °K owing to the low E_g, and little work has been performed on this material (Mukai and Irita, 1967). GaAs, although suitable for use at room temperature, has a poor noise performance. GaAs diodes also suffer some degradation if driven into the thermal-injection region, but this is unlikely to occur under normal amplifier conditions. GaSb offers the lowest noise factor, but with a number of limitations, and therefore Ge probably provides the best overall performance.

7.4.4 Frequency ranges

T.D.A.s are readily obtainable in Britain to a wide variety of specifications between 0·5 and 7 GHz, but there are few suitable diodes for X band frequencies and above, and package inductance and capacitance make design difficult. Above J band, Burrus and Trambarulo (1961) have reported on t.d.a.s operating at frequencies up to 90 GHz with bonded-contact Ge tunnel diodes made in reduced-height waveguide within the amplifier.

7.4.5 Overload characteristics

The peak and valley voltages of a tunnel diode are constant for a given material; so any increase in power handling must be obtained by increasing the peak current. But, as R_n is inversely proportional to peak current, the choice is limited by circuit-design considerations. The majority of t.d.a.s use germanium diodes with a peak current I_p of the order of 2 mA. The input level producing 1 dB gain compression in a single-stage germanium t.d.a. is approximately

$$1 \text{ dB compression point} = -(2G + 20I_p) \text{ dBm} \qquad (7.10)$$

where G is in decibels and I_p is in milliamperes. The 1 dB compression point can be increased by up to 10 dB on selecting an appropriate bias voltage somewhat higher than the I/V inflection point. Thus, a 15 dB-gain amplifier using a 2 mA diode has a 1 dB compression point of -40 dBm when biased for maximum gain and bandwidth at the I/V

inflection point. This can be increased to -33 dBm on increasing the bias voltage. Obtaining the same gain by cascading two low-gain amplifiers gives a further increase in the 1 dB compression point.

One example of obtaining a greater dynamic range without sacrifice in noise factor has been the use of a low-noise Ge tunnel diode followed by a higher-power (and higher-noise) GaAs diode in cascade; this arrangement has been shown by Steinhoff and Stertzer (1964) to result in a 15 dB improvement in dynamic range.

T.D.A.s have also been made using higher-current diodes; a 1 dB compression point of -25 dBm using a 22 mA GaAs diode has been reported. However, package inductance is the main obstacle to the use of high-current tunnel diodes owing to the low values of f_{x0} obtained when R_n is small. An alternative approach by Lee (1967) makes use of two germanium diodes in a push–pull arrangement, which both improves the power-handling capabilities and reduces the intermodulation response of the amplifier.

7.4.6 Distortion

Intermodulation distortion remains very small (signal/distortion > 70 dB) unless the magnitude of the interfering signal is sufficiently large to cause severe gain compression of the amplifier.

A.M.–p.m. conversion can result at large signal levels, but again this effect is small. For example, 0·2 deg/dB a.m.–p.m. conversion occurs at an input level of -20 dBm for a 2 mA germanium diode (Easter, 1965).

7.4.7 Burnout

Permissible c.w. dissipations are of the order of a few tens of milliwatts, but the spike-burnout levels are more difficult to determine, and more extensive tests are required to give quantitative results. American data suggest single-pulse-burnout levels of about 1 erg, but it is difficult to correlate this test with practical conditions and it is significant that the first t.r.-cell–varactor limiters were originally developed in order to protect the t.d.a. Therefore the burnout level of tunnel diodes is probably similar to that of mixer diodes at about 0·1–0·5 erg for repetitive pulsing. There is also some evidence to suggest that, from this aspect, Ge diodes are superior to either GaAs or GaSb devices, but the main factor is probably the design of the junction element.

7.4.8 Temperature

Ge and GaSb have useful operating temperatures of -55 to $100\,^\circ$C, whereas GaAs can be used up to $125\,^\circ$C. This is not a severe restriction in military applications where temperature stabilisation would be employed. However, the peak current of GaSb diodes is more temperature-sensitive and this tends to nullify the slight advantage in noise factor of this material.

7.5 Future trends

The main promise of the t.d.a. is its wide bandwidth, the main difficulty is the low saturation power of the present diodes. Assuming that the package capacitance could be eliminated, Table 4 of Section 7.4.2 could be extended as follows for 20 dB gain:

Table 6

	Bandwidth with single tuned circuit GHz	Maximum theoretical bandwidth GHz
Including C_s	1	6·5
Neglecting C_s	2·3	15·7

So it is clear that improved bandwidth can be obtained by the use of unencapsulated diodes which could be used with other integrated-circuit elements. The decrease of diode inductance thus obtained would also improve circuit stability and enable higher-current tunnel diodes to be used, giving the better power handling required.

Thin-film circuits offer a good initial approach, and Okean (1966b) has made a start in this direction on a 4 GHz amplifier. This reflection amplifier has a thin-film tuning inductance, a bypass capacitor, and an

130

LCR stabilising circuit formed on a small substrate of sapphire. The diode is made by a bonded contact on a small die of germanium, coated with epoxy resin, and subsequently mounted on the substrate.

A considerable improvement in noise factor appears to be unlikely. The noise depends mainly on the noise constant N_s, which is proportional to $I_0 R_n$. The use of higher- or lower-peak-point-current diodes will not alter the noise measure since R_n is inversely proportional to I_0, resulting in a constant value for the product $I_0 R_n$. On employing different semiconductors or doping techniques, N_s may be altered, the effect being essentially to modify the voltage scale of the characteristic. Lower peak-point and valley-point voltages result in a lower N_s, but also a lower dynamic range. The use of GaSb gives a noise factor approximately 0·75 dB lower than that of Ge, but this semiconductor has a less well developed technology and tunnel diodes made from it are reported to be more temperature-sensitive and fragile.

7.6 Summary

The tunnel-diode amplifier is at present used mainly as a simple moderately-low-noise preamplifier for microwave receivers, in cases where the additional complexity of the parametric amplifier is not justified or where it is desirable for the overall noise factor to be dependent on the noise performance of one unit rather than two or more as with a mixer–i.f. chain. The t.d.a. has some power-output limitations, but all low-noise amplifiers have dynamic signal-range restrictions for some system applications, and the t.d.a. has a range of about 60 dB (17 dB gain and 20 MHz bandwidth) and this can be adequate for many applications.

A synthesis procedure is available which enables a stable t.d.a. to be designed with a specified frequency response, taking into account all the circuit elements.

It is possible that, with the advent of better microwave transistors and mixers, the t.d.a. will become obsolete. However, it is evident that some of the potential advantages of the t.d.a. have yet to be realised.

The noise factor of a tunnel diode is governed by the noise constant, cutoff frequency and resistance ratio, and for a germanium device this puts a theoretical lower limit of about 3·5 dB on the noise factor. The design of these devices is now well understood and noise factors of this order have been achieved at L band and 5·5 dB in X band frequencies.

Operation at a higher current, using a higher-peak-point-current diode, increases the dynamic range but increases the stability problems associated with the higher negative conductance. An alternative approach is to cascade a low-noise Ge t.d.a. with a higher-power GaAs t.d.a.

The stability problems associated with microwave tunnel-diode circuits arise from having a low self-resonant frequency f_{x0}. This frequency is determined principally by the inductance of the diode housing L_s. Further reduction of L_s is very difficult, using diodes in discrete form, but the difficulty would be largely avoided by adopting microcircuit form. Building the amplifier in integrated or thin-film-circuit form may make it possible to reduce stability problems and achieve a higher dynamic range. It is also likely to simplify the design of the associated filter circuits.

Further work on tunnel diodes is needed. At present most alloyed diodes are of etched-mesa construction, somewhat fragile at higher frequencies, and incompatible with integrated circuits. Some progress is being made with planar techniques, the main difficulty being in defining the minute junction area required. However, it now seems feasible to consider an integrated t.d.a. which would offer improved bandwidth and dynamic range. A discussion of microwave amplifiers employing tunnel-diode devices has recently been published by Okean (1967).

8 MICROWAVE LOW-NOISE TRANSISTOR AMPLIFIERS

8.1 Introduction

Following the discovery of the transistor in 1948, considerable effort has been devoted to improving its high-frequency performance. Although other transistor structures have been explored, the main approach has been to extend the bipolar transistor by means of improved technology. The change from point-contact, to alloyed, to diffused, to epitaxial and to planar techniques has steadily moved transistor operation into the microwave region until a 6 GHz amplifier is now feasible. It would appear that further technological development will slowly extend transistor operation towards X band frequencies. The transistor amplifier is basically simple with modest power supplies, no pumps or circulators, and a reasonable noise factor. Commercially available germanium transistors already have noise factors of 4 dB at 2·25 GHz, and silicon transistors have noise factors of 5 dB at 2 GHz. In the laboratory, even better results are quoted, with germanium transistors, ranging from 2·6 dB at 1·5 GHz to 4·5 dB at 3 GHz and 8 dB at 5·5 GHz. Silicon transistors have noise factors of 2·5 dB at 1 GHz and 3·6 dB at 2 GHz.

The following sections elaborate on these points and consider the future possibilities.

8.2 Sources of noise in bipolar transistors

As for all electronic devices, the sensitivity of the transistor as an amplifier is limited by the effects of internally generated noise. This has a number of distinct forms: thermal, shot, generation–recombination and flicker noise.

8.2.1 Thermal noise

Thermal noise is caused by thermal agitation of the current carriers in the bulk material of the transistor, giving them a random motion. Associated with the ohmic resistance R of the bulk material of the transistor there is a noise current i_n of which the mean square is

$$\overline{i_n^2} = 4kTG\,df \tag{8.1}$$

where $G = 1/R$, and df represents an infinitesimally small frequency range. Thermal, 'white', or Johnson, noise consists of the summation of short random current pulses and is therefore uniform over the whole frequency band. It is correlated with the real part of the resistance of the transistor and it thus depends upon the base and collector resistances and any contact resistances. There is some thermal noise from the generator resistance R_g, which also includes biasing resistors or feed-back resistance in parallel with the emitter.

8.2.2 Shot noise

Under bias conditions, the transistor is no longer in thermal equilibrium and additional noise arises from the flow of electron and hole currents. This constitutes 'shot' noise, which has a white spectrum at frequencies which are low in comparison with the reciprocal of the carrier transit time across the depletion layer. Shot noise is current-dependent, the mean-square shot-noise component of a current I being

$$\overline{i_n^2} = 2qI\,df \tag{8.2}$$

The current carriers which surmount the potential barrier at the emitter and are injected into the base as minority carriers move towards the collector mainly by diffusion. The movement is accelerated when an internal drift field is present, but the moving carriers still undergo a large number of collisions with the crystal lattice and with the majority carriers. Hence fluctuations in the diffusion process form one of the main sources of transistor shot noise. However, not all the injected carriers reach the collector. Some recombine with majority carriers and new pairs of carriers are also generated by thermal agitation of the crystal lattice. Fluctuations in the generation–recombination process of minority charge carriers is the second fundamental source of transistor shot noise.

8.2.3 Generation–recombination noise

Fluctuations in the number of free majority carriers in the emitter, base and collector regions, owing to generation and recombination processes, constitute possible noise sources in transistors. This so-called semiconductor current noise (or g.r. noise) is strongest in intrinsic and nearly intrinsic material. In the extrinsic regions of emitter, base and collector of practical transistors the majority carrier density is determined by the density of impurity atoms, which are completely ionised at room temperature. Therefore the contribution of this type of noise is negligible under normal conditions.

8.2.4 Flicker noise

Another type of noise, which is unimportant at high frequencies, is flicker noise, which is caused by the variation of leakage current and surface recombination velocity with surface properties. Flicker noise is proportional to $1/f$ down to very low frequencies and is generally negligible above 2 kHz.

8.3 High-frequency noise performance of bipolar transistors

The noise of a transistor is relative to the noise at the input, i.e. relative to noise in the generator, which is therefore the minimum possible noise. In practice, the minimum transistor noise factor can approach 1 dB at a current level of about 1·0 mA. The gain of a transistor falls in proportion to f/f_T, where f_T is the total unity-gain frequency. Therefore the contribution of the input noise (at the emitter) to the total output noise drops as the gain drops, because the input noise is being amplified through the transistor. The collector noise, which is not being amplified, becomes predominant at high frequencies and increases at 6 dB per octave relative to the input noise, because the latter is falling at the same rate as the gain falls. The corner frequency of the noise-factor/frequency curve, i.e. the frequency above which the noise increases at the rate of 6 dB per octave, is approximately equal to $f_\alpha(\beta_0)^{-\frac{1}{2}}$. Therefore, in order to remain in the flat or white-noise region up to as high a frequency as possible, β_0 should not be too large. Taking a typical value of β_0 to be 50, for a corner frequency of 1 GHz, f_α would have to be 7 GHz.

The noise performance of a transistor is generally assumed to be given by Nielsen's formula (Nielsen, 1957). This formula predicts the same noise factor for the common-emitter and common-base connections, namely

$$F = 1 + \frac{r_b'}{R_g} + \frac{r_e}{2R_g} + \frac{(r_b'+r_e+R_g)^2}{2\alpha_0 R_g r_e}\left\{\frac{1}{\beta_0}+\left(\frac{f}{f_\alpha}\right)^2+\frac{I_{cb0}}{I_e}\right\} \qquad (8.3)$$

It is important, therefore, to have a low base resistance r_b', emitter resistance r_e, and collector spreading resistance r_{sc} (which is not included in the formula) as well as the optimum generator resistance R_g. The alpha cutoff frequency f_α should be as large as possible. Nielsen's formula predicts less high-frequency noise for the common-collector connection, but this is not a usual mode of operation.

The agreement between theory and experiment for noise in germanium and silicon v.h.f. transistors becomes poorer above 1 GHz. This deterioration seems to be due partly to the effect of transistor-header parasitics. Experimental evidence (Granberry and Policky, 1965) has shown also that microwave-transistor noise is lower in the common-emitter than in the common-base connection.

Fukui (1966) has considered the noise performance of microwave transistors. He finds that the high-frequency noise factor is given, approximately, by

$$F \simeq 1 + h\left\{1 + \left(1 + \frac{2}{h}\right)^{\frac{1}{2}}\right\} \qquad (8.4)$$

where

$$h = \frac{qI_c r_b'}{kT}\left(\frac{f}{f_T}\right)^2$$

so that F is a function of I_c. Eqn. 8.4 is a good approximation if the transistor has sufficiently good d.c. characteristics, i.e. if

$$h_{fe} > 10(f_T/f)^2$$

As I_c decreases, the shot noise caused by fluctuations in the collector and base currents decreases; but f_T also decreases and in some cases the base resistance increases. Hence there is a minimum F at a particular collector current, which may be found by differentiation of eqn. 8.4. For a constant value of r_b', the value of I_c for F_{min} is

$$I_{cmin} = 0 \cdot 16 f_T C_e \qquad (8.5)$$

where I_c is in milliamperes, f_T in gigahertz and C_e is in picofarads. C_e is the sum of the emitter–base transition-region capacitance and the base–emitter-header stray capacitance.

Feedback effects due to intrinsic and extrinsic collector-transition capacitances, collector-diffusion capacitance and collector–base-header stray capacitances reduce the noise factor, but collector-body resistance and other lossy elements associated with the header tend to cancel this reduction. The feedback effect also reduces the gain, and therefore may not improve the overall noise performance of a microwave transistor, since both noise factor and gain are important for optimum performance.

8.4 Design criteria for low-noise high-frequency transistors

Inspection of eqn. 8.3 reveals that both r_b' and r_e should be as small as possible. The base resistance is mainly determined by the geometry of the transistor and the sheet resistance of the base region. The emitter–base spacing should be as small as possible and the emitter stripes should be as long as is compatible with the capacitance requirements, which influence f_T. The sheet resistance in the base region should be as low as is compatible with the requirements of emitter efficiency, which influences α_0, and emitter capacitance, which influences f_T. A shallow diffusion has the advantage that the total number of impurities in the base can be much greater for a given base width and f_T, and this will give a smaller base resistance. The base resistance may also be decreased by increasing the emitter current I_e because of conductivity-modulation effects in the base at high current densities. An increase of emitter current is an advantage in the final term of eqn. 8.3 and also reduces the emitter resistance. The latter may be reduced by increasing the emitter area, but this would increase the value of the emitter transition-region capacitance which should be small for good high-frequency performance.

The common-base current gain α_0 should be near unity. This is influenced by the emitter efficiency, the collector efficiency and the base transport factor. The latter should be excellent for a microwave transistor and the collector efficiency is generally close to unity. A heavily doped emitter, with width greater than a minority-carrier diffusion length in the emitter, is required for good emitter efficiency. This requirement is especially severe at the edges of the emitter. The fall off of emitter efficiency with increasing I_e causes increased recombination noise.

The common-emitter current gain β_0 should be large, which, of course, is compatible with an α_0 of nearly unity, but not too large since it influences the corner frequency of the noise-factor/frequency curve.

The common-base cut-off frequency f_α should be as large as possible. It has been shown (te Winkel, 1959) that

$$f_\alpha \simeq f_T \left\{ 1 \cdot 21 + 0 \cdot 09 \log_e \left(\frac{N_e}{N_c} \right) \right\} \tag{8.6}$$

where N_e and N_c are the impurity concentrations in the base next to the emitter and collector depletion layers, respectively. To get a high ratio of $f_\alpha : f_T$, it is important to have a large impurity gradient in the base, i.e. a large drift field. According to Cooke (1965), presently available germanium transistors have $f_\alpha/f_T \simeq 2$, whereas silicon transistors have $f_\alpha/f_T \simeq 1 \cdot 2$. Presumably, a silicon transistor has a low f_α/f_T ratio because it has a diffused emitter, which reduces the overall drift field in the base, whereas germanium units have alloyed emitters. Under high-injection-level conditions the built-in drift field in the base is reduced, and this will lower the f_α/f_T ratio.

The collector–base leakage current I_{cbo} should be as low as possible, since this contributes to shot noise, and the collector spreading resistance, which contributes to thermal noise, should also be small. The latter implies the use of epitaxial collector material.

The encapsulation used for microwave transistors can have a large effect on the final parameters of the device. The standard TO–18 encapsulation is unsuitable owing to the high internal-lead inductances, the high interelectrode capacitance and high series resistances. These last occur, at microwave frequencies, where the unplated Kovar lead wires pass through the glass seal.

The preferred construction for small-signal microwave-transistor amplifiers at present consists of stripline circuitry produced on low-loss printed-circuit board. This suggests a device outline with three coplanar electrodes. Plate 19 shows such an encapsulation which has small parasitic reactances.

The future amplifier may well consist of a hybrid integrated circuit in which the passive components are deposited by thin-film techniques on a ceramic substrate. The ideal transistor outline would then be a flip-chip, beam-lead or 'l.i.d.' package. The l.i.d. is also shown in Plate 19.

8.5 General aspects of amplifier-circuit design

As mentioned earlier, the best performances so far have been achieved with germanium transistors.

The circuit designer has the choice of two configurations: common-emitter or common-base. Although a common-base connection does, in general, give a greater power gain at high frequencies, it is only unconditionally stable at frequencies above about f_T. However, the common-emitter circuit is unconditionally stable above $f = (r_e'/2r_b')f_T$, where r_e' = extrinsic emitter resistance. Common-emitter connection is thus usually employed for frequencies below the f_T of the best available transistors and common-base connection is only resorted to for higher frequencies.

There are two approaches to the design of the amplifier circuit:

(*a*) An extension of low-frequency techniques using lumped components. The upper frequency limit for this approach appears to be 1–3 GHz at present, depending on the nature and quality of the components and on the method of construction.

Some work is currently being carried out on the design of lumped components for use at frequencies up to X band. Capacitors, inductors and resistors have been fabricated using integrated-circuit techniques.

The reactive elements have small losses compared with those of semi-conductor devices and this should enable true lumped circuits to be made at frequencies up to X band. The main advantages to be expected from this approach are greater bandwidth and smaller size.

(*b*) The use of distributed circuits, derived from conventional microwave components, designed to be electrically and mechanically compatible with transistors. The best known example is the balanced amplifier described by Engelbrecht (1965) and Kurokawa (1965). The main features of this circuit are described in Section 8.10.

The detailed design of transistor amplifiers at microwave frequencies will not be dealt with here since this is covered in the literature (Engelbrecht, 1965; Kurokawa, 1965; Linvill, 1961; Anderson, 1967 and Gelnovatch, 1967). However, the characteristics to be expected from transistor amplifiers are discussed below and the factors governing these characteristics are mentioned.

8.6 Transistor-amplifier characteristics

8.6.1 Gain and bandwidth

The maximum available power gain G from a transistor under neutralised and matched input and output conditions at a frequency f is given by

$$G = \left(\frac{f_{max}}{f}\right)^2 \tag{8.7}$$

where f_{max} is the frequency at which $G = 1$ and is given by the approximation

$$f_{max} \simeq \sqrt{\left(\frac{f_T}{8\pi r_b' C_c}\right)} \tag{8.8}$$

In practice, f_{max} is generally two or three times f_T.

It follows from the above that $G^{\frac{1}{2}}B$ is equal to f_{max}. However, this applies to the active transistor element alone, not to a complete amplifier. For practical amplifiers the achievable gain–bandwidth product is somewhat less; the best results reported to date include 10 dB gain per stage at L band frequencies with an octave bandwidth to 3 dB gain per stage at C band with a 1 GHz bandwidth.

8.6.2 Noise

Nielsen's formula (eqn. 8.3) is generally used to predict amplifier
noise performance. This is, however, a simplified expression which
only applies to the susceptance-matched input and output
conditions.

Eqn. 8.3 can be differentiated with respect to I_e (both r_e and f_α are
functions of I_e) to find the optimum I_e for minimum noise. It is also
possible to differentiate with respect to R_g to find the optimum
generator resistance.

In practice, an I_e of about 1–3 mA is found to give the lowest noise
factor and the optimum R_g is, most conveniently, a few tens of ohms.
The conditions for minimum noise factor are not necessarily the same
as the conditions for maximum power gain and it is often necessary to
choose a compromise between the two.

The best noise performance achieved to date is probably that
reported by Engelbrecht (1965), who claimed a noise factor as low as
3 dB in the 1–2 GHz range and a gain of 40 dB for a 4-stage amplifier.

8.6.3 Phase characteristics

The ideal phase characteristic is one in which the phase varies
linearly with frequency. The active transistor element itself can be
expected to give a departure from linearity of less than $1°$ even for
1 GHz-bandwidth amplifiers. The transistor encapsulation, however,
can have a large influence, since the parasitic reactances have to be
tuned out, and this results in a narrowband amplifier. In order to
achieve wide bandwidths with multistage amplifiers, staggered tuning
is often used and this gives a very poor phase response.

The balanced amplifier described by Engelbrecht (1965) can be
made to give a good phase response since multistage amplifiers can be
made without the use of stagger-tuned interstage transformers.
A phase deviation from linearity of less than $1°$ over a 20 % bandwidth
has been claimed for this circuit.

8.6.4 Dynamic range and intermodulation

The smallest signal which can be detected is set by the available noise power at the input.

$$\text{Available noise power} = kT\,\Delta f\,F \qquad (8.9)$$

where $\qquad\Delta f$ = bandwidth

and $\qquad F$ = amplifier noise factor (ratio)

For an amplifier at room temperature with a 1 GHz bandwidth and a 4 dB noise factor, the available noise power at the input is -80 dBm. If the bandwidth is cut to 100 MHz, the noise power drops to -90 dBm.

The signal amplitude which can be handled is limited by saturation in the last stage. In practice, gain compression is likely to set in at about 1 mW of output power.

Thus the dynamic range would be from -80 (or -90) dBm at the input to 0 dBm at the output. This would correspond to an input dynamic range of from -80 (or -90) dBm to $-G$ dBm, where G is the amplifier power gain.

Recovery after overload is mainly a matter of circuit design, in that any CR time constants in the coupling and decoupling circuits should be short. If the transistors bottom heavily under overload conditions, the stored charge in the transistors may take some time to decay, but this should be about two microseconds at the most and can also be minimised by careful circuit design.

Consideration of the dimensions of the device suggests that the burnout levels will be a few ergs or a few tens of ergs, an order of magnitude better than that of conventional point-contact mixer diodes.

Large intermodulation products will, of course, be caused by the nonlinearity in the overall gain characteristic when compression occurs at large signal levels. Even at modest signal levels intermodulation can be a problem. The main causes are the nonlinearity of the emitter–base diodes and the variation of gain (or f_T) with emitter current. These problems can be eased by feeding from high source impedances, running at relatively high emitter currents, and using transistors designed to have a very flat f_T/emitter-current characteristic. In practice, a compromise has to be chosen between the intermodulation, gain and noise requirements.

Typical results so far reported have claimed intermodulation products 45–50 dB down on an output signal of about 0·5 mW.

8.6.5 Environmental limits

Transistors will operate satisfactorily at temperatures from about $-55\,°C$ to $+85\,°C$ (for germanium) or $+175\,°C$ (for silicon) and the encapsulations will normally withstand all humidity conditions.

There is a problem in using transistors in the presence of high-energy radiation. This has been the cause of several Earth-satellite failures and the problem is being vigorously tackled in the USA.

8.6.6 Life

Very little information on this aspect has been reported. The reason is probably that there is no life problem and that a microwave-transistor amplifier has a life similar to that of a transistor used at lower frequences; i.e. many tens of thousands of hours.

8.6.7 Ancillary equipment

For the majority of applications, no ancillary equipment (other than power supplies) is required. In some instances it may be necessary to precede the amplifier with a limiter to protect it from high-power input signals.

8.6.8 Size and weight

Apart from the mixer, the transistor amplifier is probably the smallest r.f. input stage. A typical amplifier would weigh a few ounces and occupy a few cubic inches.

The latest trends towards hybrid integrated circuits on ceramic substrates, perhaps in conjunction with lumped passive components, make it possible to produce an amplifier on a thin ceramic substrate about 1 in square. The overall size and weight are then determined almost entirely by the connectors used.

8.7 Other devices

8.7.1 Unipolar transistors

Recent contributions to the theory of noise in unipolar transistors have been made by Bruncke and van der Ziel (1966) for the junction-gate field-effect transistor (f.e.t.) and by Sah *et al.* (1966) for the insulated-gate field-effect transistor (m.o.s.t.). The main source of noise in both of these devices is the thermal noise of the conducting channel, which is practically independent of frequency, except near the cutoff frequency. The equivalent noise resistance is given by $R_n = \frac{1}{2}g_m^{-1}$ for the f.e.t., and $R_n = \frac{2}{3}g_m^{-1}$ for the m.o.s.t. The factor $\frac{1}{2}$ or $\frac{2}{3}$ arises because of the difference of the channel shape in the two devices. In the case of an m.o.s.t. on a conducting substrate, there is an additional noise contribution from the effective transconductance between the substrate and the channel, which increases the value of R_n. This effect becomes appreciable when the substrate doping level exceeds 10^{16} atoms/cm^3, and values of R_n up to 3 or 4 times g_m^{-1} have been reported (Mavor and Reed, 1967).

The internal source resistance of the device constitutes an unbypassed resistance, the thermal noise of which is amplified by the transistor action. It is important, therefore, to have the source resistance as small as possible. The external generator resistance also contributes to the overall noise, and there is an optimum generator resistance for minimum noise factor.

At high frequencies, the drain–gate feedback capacitance causes thermal noise to be fed back to the gate and amplified. However, since the signal is also fed back, the signal/noise ratio is, to a first approximation, unchanged. At high frequencies, the capacitive coupling between the channel and the gate gives rise to a capacitance noise current; in other words the input conductance g_{11} gives almost full thermal noise, dependent on the square of the frequency.

The gate-leakage current of the f.e.t. generates shot noise, which is independent of frequency, but little correlation exists between this noise and the thermal noise of the channel. The carrier generation–recombination phenomena in the gate space-charge region near the surface generate noise inversely proportional to the square of the

144

frequency. Flicker noise, due to surface effects, inversely proportional to the frequency is also present. Owing to the very small gate-leakage current, which is negligible in the m.o.s.t., and the frequency dependence of these other noise sources, they may be neglected at frequencies above about 10 kHz.

Bruncke and van der Ziel (1966) have given a minimum noise factor F_{min} for f.e.t. amplifiers at high frequency

$$F_{min} \approx 1 + 2(R_n g_{11})^{\frac{1}{2}} \qquad (8.10)$$

and predict a value for the noise factor somewhat above the cutoff frequency, of the order of 7 dB.

Experimentally, it is found that the noise parameters of an f.e.t. approach limiting values at the onset of drain-current saturation which are maintained well into the saturation region of the characteristic. The noise level of good units is comparable to that of the best low-noise bipolar transistors, but as yet the high-frequency performance is not as good. A currently available m.o.s. tetrode has a noise factor of 4 dB at 500 MHz and a triode version should be slightly better than this. The theoretical minimum noise factor for a tuned and and matched amplifier should be of the order of 3·7 dB at or below f_T for both bipolar and unipolar transistors. The high-frequency thermal-noise characteristics of the f.e.t. and m.o.s.t. are both somewhat better than those of a vacuum tube with equal transconductance.

8.7.2 Optotransistor or transluxor

In the optotransistor, or transluxor, the charge is transferred, in effect, by a photon resulting from the combination of a hole and electron at the emitter, which then decomposes into an electron and hole in the depletion layer of the collector. In operation, the emitter junction is forward-biased and the collector junction reverse-biased, as in a transistor. The main advantage of the optotransistor is that it enables a semiconductor with a very low minority-carrier lifetime to be made into a transistor with a relatively large base width; in fact a low carrier lifetime is necessary because the time taken for the carriers to recombine (i.e. the delay before they emit light) is the most important limiting time constant of the device. Other advantages are that the base resistance can be made very small ($1\ \Omega$), giving a considerable reduction in thermal noise, and the signal transport from emitter to collector is at the speed of light in the material.

Gallium-arsenide optotransistors, employing a light-emitting gallium-arsenide p–n junction for the emitter and an n type gallium arsenide and p type germanium heterojunction for the collector, have been made. Theoretically (Yu, 1963) these devices could operate at X band frequencies with high gain and reasonable noise factors. However, present-day optotransistors are limited by the difficult technology involved.

8.7.3 Metal-base transistors

Of all the hot-electron transistors with thin metal bases that have been proposed, only the semiconductor–metal–semiconductor arrangement appears to show any promise of a significant increase in operating frequency over those of bipolar transistors (Sze and Gummel, 1966). This device operates by injection of electrons over a Schottky barrier from a semiconductor emitter through a thin metal film and over a second Schottky barrier which acts as the collector. Low-frequency devices have been fabricated using Si–Au–Ge and Si–Ag–Ge layers. Microwave operation would appear to be attractive because of the small base resistance and negligible base transit time; however, because of fundamental physical limitations, the common-base current gain is low ($\alpha = 0{\cdot}3$). At 10 GHz it is calculated that this transistor would provide a power gain of 20–30 dB with a minimum noise factor of 8 dB. The greatest unsolved technological problem is the deposition of device-quality semiconductor materials on thin metal films, and to date there have been no reports of epitaxial semiconductor deposition on metals such as Au and Ag which have a long mean free path for electrons.

8.8 Future trends

8.8.1 Transistors

Germanium transistors have superior high-frequency noise performance compared with silicon ones at the present time, owing to the greater f_α/f_T ratio and the lower base resistance of a germanium p–n–p transistor compared with a silicon n–p–n transistor. This is because of the greater electron mobility in germanium compared with hole mobility in silicon. However, the technology of silicon devices is on a much firmer foundation than that of germanium; the performance gap between silicon and germanium transistors is closing rapidly and 10 GHz operation may finally be achieved with silicon transistors. Plate 20 shows a small-signal, low-noise, silicon-planar microwave transistor with 1 μm-wide emitter fingers. These dimensions represent the limit which may be achieved at the present time with conventional silicon-planar technology. Silicon transistors offer higher gain and in some instances can provide superior overall receiver noise factors even though the individual transistor noise performance is worse than that of a germanium transistor. In these cases the contribution by the following stages is more effectively reduced.

For a very good low-noise high-frequency transistor having $f_T = 4$ GHz, $f_\alpha = 7$ GHz, $f_{max} = 10$ GHz, $C_e = 0\cdot5$ pF, and $r_b' = 20\,\Omega$, eqns. 8.4 and 8.5 predict almost the same noise performance as that mentioned for the practical devices in Section 8.1. The noise performance predicted is in fact $3\cdot5$ dB at $1\cdot5$ GHz and 7 dB at 3 GHz. If r_b' were reduced to 10 Ω, the predicted noise would be $2\cdot3$ dB at $1\cdot5$ GHz, 4 dB at 3 GHz and 11 dB at 8 GHz. This performance is only slightly better than the laboratory results mentioned in Section 8.1 and would be close to the feasible limit attainable with the present technology, and so new processes will have to be evolved for further improvement. If the critical dimensions can be reduced by electron-beam techniques, either to improve the resolution of photolithography (Thornley and Sun, 1965) or to decompose an oxysilane compound in certain precisely determined regions, thereby producing high-resolution SiO_2 patterns (Ford *et al.*, 1966), the high-frequency noise performance of transistors should be further improved. Additional improvements in r_b' and the f_α/f_T ratio should be possible using ion implantation (Kerr and Large, 1966), which would also improve the high-frequency noise factor. A lower r_b' should be obtained with a silicon p–n–p transistor, rather than the more common n–p–n, and also with a p–n–p gallium-arsenide transistor if gallium-arsenide-transistor technology improves to equal that of silicon. In this respect, ion implantation might prove to be a much better technique than diffusion for compound semiconductors.

Finally, improved versions of transluxors or metal-base transistors could give a very low base resistance and therefore a very good high-frequency noise factor.

8.8.2 Circuits

The most promising circuit design for the immediate future would appear to be the balanced circuit, using two transistors per stage. This can be conveniently constructed using printed-circuit or thin-film techniques.

It is possible to produce monolithic versions of such circuits using a high-resistivity semiconductor wafer as the substrate with the transistors diffused into the substrate and the transmission-line components deposited on the upper surface. However, since the size of distributed components based on transmission lines is governed by the wavelength, the wafer of semiconductor material required is rather large, e.g. 1 in square at L band frequencies, $\frac{1}{4}$ in square at C band frequencies, $\frac{1}{8}$ in square at X band frequencies. This therefore seems an unlikely approach, certainly at frequencies below X band.

The rapid progress now being made with conventional integrated circuits may make possible monolithic integrated microwave amplifiers with lumped components, since the problems associated with lead inductances can be reduced to negligible proportions. However, at present, the stray capacitance of a conventional diffused resistor is too large and the Q factor of a diffused capacitor is too low for these to be used at microwave frequencies. These passive components would need to be deposited by thin-film techniques onto a thick oxide layer over the semiconductor substrate.

8.9 Conclusions

The rapid development of planar technology has extended the use of transistors into the microwave region. Promising amplifiers have been made with germanium transistors with a gain of 12 dB and a noise factor of 7·3 dB at 4 GHz. Further improvements in technology will give greater control over the delineation of the junction and contacts which will probably eventually provide equivalent performance in silicon and extend the operating frequency to X band.

8.10 Balanced-transistor amplifier

This is shown diagrammatically in Fig. 36. It consists of a pair of matched transistors each with its own input and output matching networks. The two inputs are combined via one 3 dB coupler and the two outputs via a second 3 dB coupler.

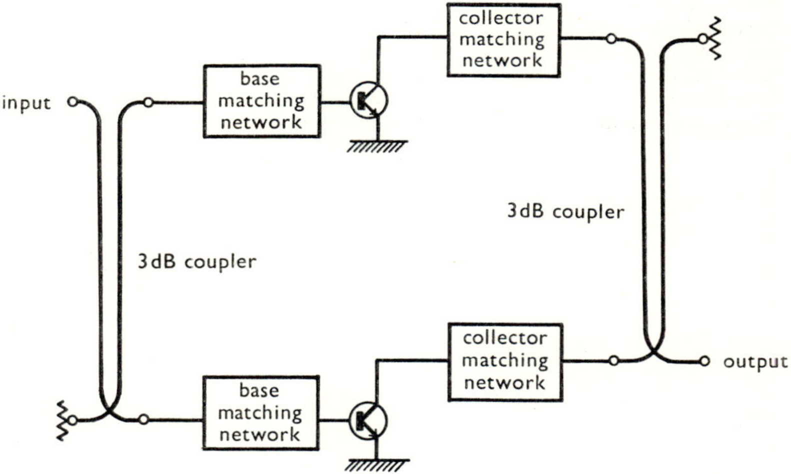

Fig. 36   Balanced-transistor amplifier

This arrangement has the following advantages over the single-transistor amplifier stage:

(*a*) The match at the input and output of each balanced stage can be maintained over a wide frequency range, even if the input and output match at the individual transistors deteriorates. Any reflections from the two transistors (provided that these are similar) are dissipated in the loads connected to the fourth ports of the 3 dB couplers. This enables stages to be cascaded with fewer interaction problems.

(*b*) A low v.s.w.r. can be maintained at the input even though the source impedance presented to the transistor is that required for optimum noise performance.

(*c*) The signal level at which saturation sets in will be doubled and this will reduce intermodulation.

The best performances reported have been obtained with this type of amplifier and include the following for 4-stage germanium-transistor amplifiers:

	March 1965	March 1966
3 dB bandwidth (b.w.)	0·65–1·7 GHz	3·4–4·4 GHz
v.s.w.r. over above range	< 2:1	< 2:1
Gain	20 dB±0·2 dB (20 % b.w.)	12 dB±0·5 dB (15 % b.w.)
Reverse isolation	50 dB (20 % b.w.)	45 dB (15 % b.w.)
Phase deviation from linear	< ±1° (20 % b.w.)	—
Input v.s.w.r.	< 1·1 (20 % b.w.)	< 1·3 (15 % b.w.)
Output v.s.w.r.	< 1·1 (20 % b.w.)	< 1·3 (15 % b.w.)
Noise factor	< 6 dB (now nearer 3 dB)	7·3 dB
Third-order intermodulation	−50 dBm output at $2f_A - f_B$ and $2f_B - f_A$, where output at f_A and f_B is −3 dBm	

SYMBOLS AND ABBREVIATIONS

α	loss factor of transmission line
α	fraction of cold loss of t.w.t. which affects the gain
α	idler loading ratio of parametric amplifier
α_0	common-base current gain of transistor
β	magnetic moment of electron
β, β_i, β_s and β_p	phase constants: general, idler, signal and pump frequencies
β_0	low-frequency current sensitivity of detector
β_0	common-emitter current gain of transistor
γ	gyromagnetic ratio
γ	capacitance-variation coefficient of varactor diode
γ	electron beam-parameter of travelling-wave tube (usually used in the combination γa)
ϵ	relative permittivity
ϵ	thermal emissivity
λ	wavelength
λ_g	guide wavelength
ϕ	contact potential
η	filling factor
τ_1	spin–lattice relaxation time
τ_2	spin–spin relaxation time
ρ	resistivity
μ	carrier mobility
ω	thickness of epitaxial layer
ω	frequency (rad/s) (subscripts as f)
Γ	reflection coefficient
Π	noise parameter of a t.w.t.
Φ	self-power density spectrum of kinetic noise-voltage variations of electron beam

Ψ self-power density spectrum of noise-current variations of electron beam

Θ cross-power density spectrum between kinetic noise voltage and noise-current variations

a electron-beam radius

a radius of point contact

B bandwidth (to 3 dB points unless specified otherwise)

B_m line width of paramagnetic resonance

c velocity of light

C intermodulation constant

C_j junction capacitance

C_s stray (case) capacitance

C_{max} junction capacitance of varactor at onset of significant forward current

C_{min} junction capacitance of varactor at maximum expected reverse bias

C_0 junction capacitance at zero pump power

C_b barrier capacitance (mixer or detector diode)

C_e combined emitter–base and stray header capacitance (transistor)

d_1, d_2 slope factors of signal and idler circuits of parametric amplifier

ΔE difference between energy levels

E_g energy gap

e electronic charge

f frequency, Hz

f_1, f_2 and f_3 various specified frequencies

f_s signal frequency ⎫

f_i idler frequency ⎬ often also referred to as f_1, f_2 and f_3

f_p pump frequency ⎭

f_p^* or f_3^* optimum pump frequency of parametric amplifier

f_{I1}, f_{I2} idlers in multiple-idler parametric amplifier

f_c resistive cutoff frequency of varactor diode

f_{c0} f_c at zero bias

f_{r0} cutoff frequency of tunnel diode

f_T frequency for transistor unity total gain

f_{max}	frequency for unity available matched power gain
f_r	series self-resonant frequency of diode
Δf	frequency difference
F	noise factor
F_{min}	minimum noise factor
F_{if}	noise factor of i.f. stage of receiver
g_m	transconductance of unipolar transistor
$g(f)$	paramagnetic line shape function
G	gain
G_0	gain at low signal level
G_e	electronic gain (of a maser)
h	Planck's constant
h	Fukui's noise parameter (transistor)
i_n	noise current
H_0	applied magnetic field
I	current
I_c	collector current (transistor)
I_{cb0}	collector–base leakage current (transistor)
I_{for}	forward current (diode)
I_p	peak current (tunnel diode)
I_0	$\frac{1}{2}I_p$ (tunnel diode)
I_0	t.w.t. beam current
K	interaction impedance between electron beam and circuit of t.w.t.
k	Boltzmann's constant
L	length of maser structure
L_c	conversion loss of mixer
L_1, L_2	losses within t.w.t.
m	electron mass
M	figure of merit of video receiver
M	figure of merit of varactor
M	matrix element determining transition probability in a maser
Δn_1	equilibrium population-density differences between maser energy levels: pump on
Δn_0	as Δn_1 with pump off

Δn_s	inverted population-density difference with zero signal
n_1, n_2, etc.	population densities of energy levels
N	total population density $= n_1 + n_2 + \ldots$
N	number of stages in multistage amplifier
N	majority-carrier density in semiconductor
N_r	noise-temperature ratio of mixer diode
N_s	noise constant of tunnel diode
N_e, N_c	impurity concentrations in the base region of a transistor, adjacent to emitter and collector depletion layers respectively
P_{mn}	power flow into nonlinear reactance
$\Delta P/P$	pump-power variation
$P(f)$	output power of low-noise amplifier at specified frequency, dBm
P_m	noise power emitted per unit length by a maser crystal
P_0	power absorbed per unit length in maser structure
P_s	maximum power available per unit length of maser
P_n	noise power
q	elementary charge
Q	circuit Q factor
Q_m	magnetic Q factor—determined by rate of emission of energy by maser crystal
Q_0	Q factor determined by ohmic losses in maser structure
r_b'	base resistance (transistor)
r_e	emitter resistance (transistor)
r_e'	extrinsic emitter resistor (transistor)
r_{sc}	collector spreading resistance (transistor)
r	radius of diode junction
R_s	spreading resistance (diode)
R_g	generator resistance
R_b	barrier resistance of mixer or detector diode
R_A	equivalent noise resistance of video amplifier following a detector diode
R_n	equivalent noise resistance of unipolar transistor
R_n	negative resistance of tunnel diode
S	noise parameter of t.w.t.

154

t_v	noise-temperature ratio of detector diode
t	time
T	absolute temperature
T_m	effective noise temperature of maser crystal
T_n	effective input noise temperature of amplifier
T_0	ambient temperature of active element of amplifier
T_c	cathode temperature
T_e	system noise temperature
T^*	reference temperature ($290\,^\circ\mathrm{K}$)
v_g	group velocity
V_c	volume of maser crystal per unit length
V	bias voltage across varactor
V_0	t.w.t. beam voltage
W_s	stored energy per unit length of maser
W_{ij}	probability of stimulated transition between ith and jth levels
w_{ij}	probability of thermal transition between ith and jth levels
Z	impedance

REFERENCES

The italic figures in square brackets indicate the pages of the text in which the reference is quoted.

ADLER, R., and HRBEK, G. (1964): 'Low noise d.c. pumped cyclotron wave amplifiers', Proceedings of the 5th international congress on microwave tubes, Paris, p. 17   [*34*]

AITCHISON, C. S. (1967): 'Low noise parametric amplifiers', *Philips Tech. Rev.*, **28**, p. 204   [*80*]

AITCHISON, C. S., DAVIES, R., and PAYNE, C. D. (1966): 'A balanced micropill diode parametric amplifier' *in* 'Microwave and optical generation and amplification'. *IEE Conf. Publ.* **27**, p. 297   [*77*]

AITCHISON, C. S., DAVIES, R., and PAYNE, C. D. (1968): 'Bandwidth of a balanced micropill diode parametric amplifier', *IEEE Trans.*, **MTT-16**, p. 46   [*77*]

ALFEYEV, V. N. (1966): 'Low temperature electronics', translated from Russian (Iliffe, 1968)   [*75, 111*]

ALLEN, C. M., LIEGEY, P. R., and SALZBERG, B. (1963): '10 °K noise temperature from indium antimonide varactors at 77 °K', *Proc. Inst. Elect. Electronics Engrs.*, **51**, p. 856   [*88*]

ANDERSON, C. H.: see SABISKY, E. S.

ANDERSON, R. W. (1967): '*S*-parameter techniques for faster, more accurate network design', *Hewlett-Packard J.*, p. 13   [*140*]

ARAMS, F. R., and PEYTON, B. J. (1965): '8 mm traveling-wave maser and maser-radiometer system', *Proc. Inst. Elect. Electronics Engrs.*, **51**, p. 856   [*52*]

ARMSTRONG, J. G. (1963): Unpublished reports   [*16, 17, 20, 24, 25, 26*]

ARMSTRONG, L. D. (1962): 'Tunnel diodes for low noise microwave amplification', *Microwave J.*, **5**, p. 99   [*127*]

BALDWIN, L. D. (1961): 'Non-reciprocal parametric amplifier circuits', *Proc. Inst. Radio Engrs.*, **49**, p. 1075   [*86*]

BENNY, A. H. (1965): 'Silicon variable capacitance diodes for use at low temperatures'. Proceedings of the joint IERE-IEE symposium on microwave applications of semiconductors, London   [*88*]

BENNY, A. H., and PEARSON, J. (1965): 'A high frequency wideband idler circuit for parametric amplifiers', *Proc. Inst. Elect. Electronics Engrs.*, **53**, p. 181   [*94*]

BERGER, N. W.: see RAUB, W. E.

BLACKWELL, L. A., and KOTZEBUE, K. L. (1961): 'Semiconductor-diode parametric amplifiers' (Prentice-Hall) [*65*]

BLOEMBERGEN, N. (1956): 'Proposal for a new type of solid state maser', *Phys. Rev.*, **104**, p. 324 [*36, 39*]

BRENNER, H. E. (1967): 'Stabilization of the gain/frequency characteristics of parametric amplifiers at high input signal levels', *IEEE Trans.*, **MTT–15**, p. 290 [*82*]

BRENNER, H. E., and EISELE, K. M. (1966): 'Determination of the pump modulation factor of varactor diodes under operating conditions', *ibid.*, **ED–13**, p. 415 [*93*]

BRUNCKE, W. C., and VAN DER ZIEL, A. (1966): 'Thermal noise in junction-gate field-effect transistors', *ibid.*, **ED–13**, p. 323 [*144, 145*]

BURRUS, C. A., and TRAMBARULO, R. (1961): 'A millimetre-wave Esaki diode amplifier', *Proc. Inst. Radio Engrs.*, **49**, p. 1075 [*128*]

CHAKRABORTY, D., and COACKLEY, A. (1967): 'Characteristics of varactor diodes at low temperatures', *Radio Electronic Engr.*, **33**, p. 97 [*73*]

CHEN, F. S. (1964): 'The comb type slow wave structure for T.W.M. applications', *Bell Syst. Tech. J.*, **43**, p. 1035 [*45*]

CHEN, F. S.: see TABOR, W. J.

COACKLEY, A.: see CHAKRABORTY, D.

COOKE, H. F. (1965): 'Advances in microwave transistors and mixer diodes'. Proceedings of the joint IERE–IEE symposium on microwave applications of semiconductors, London [*138*]

COOPER, B. F. C.: see JELLEY, J. V.

CURRIE, M. R. (1958): 'A new type of low-noise electron gun for microwave tubes', *Proc. Inst. Radio Engrs.*, **46**, p. 570 [*14, 33*]

DAVIES, R.: see AITCHISON, C. S.

DEGRASSE, R. W., KOSTELNICK, J. J., and SCOVIL, H. E. D. (1961): 'The dual channel 2390 mc traveling-wave maser', *Bell Syst. Tech. J.*, **40**, p. 1117 [*56*]

DEGRASSE, R. W., SCHULZ-DUBOIS, E. O., and SCOVIL, H. E. D. (1959): 'The three-level solid state traveling-wave maser', *ibid.*, **38**, p. 305 [*43*]

DE GRUYL, J. A., OKWIT, S., and SMITH, J. G. (1965): 'Techniques for providing TWM's with wide instantaneous bandwidths', Symposium on microwave theory and techniques, Clearwater, Fla., USA [*56*]

DEJAGER, J. T. (1964): 'Maximum bandwidth performance of a non-degenerate parametric amplifier with single-tuned idler circuit', *IEEE Trans.*, **MTT–12**, p. 459 [*78*]

DE LOACH, B. C. (1964): 'A new microwave measurement technique to characterise diodes and an 800 GH_z cut-off frequency varactor at zero bias', *ibid.*, **MTT–12**, p. 15 [*91*]

DOHERTY, W. (1966): 'Precise cut-off frequency characteristics of varactor diodes', Proceedings of the international solid state circuits conference, Philadelphia [*95*]

EASTER, B. (1965): 'The performance and limitations of low-level amplifiers employing tunnel diodes', *Electronic Engng.*, **37**, p. 520 [*124, 129*]

EDRICH, J. (1966): 'Low noise parametric natural resonance amplifiers with large bandwidths', *Frequenz.*, **20**, p. 337 [*79*]

EISELE, K. M.: see BRENNER, H. E.

ELWARD, J. P.: see UENOHARA, M.

ENGELBRECHT, R. S., and KUROKAWA, K. (1965): 'A wideband low noise L-band balanced transistor amplifier', *Proc. Inst. Elect. Electronics Engrs.*, **53**, p. 237 [*140, 141*]

ESAKI, L. (1958): 'New phenomenon in narrow germanium $p–n$ junctions', *Phys. Rev.*, **109**, p. 603 [*117*]

FEHER, G.: see SCOVIL, H. E. D.

FINK, H. J., and RULISON, R. L. (1964): 'Epitaxial silicon varactors at low temperatures and microwave frequencies', *Proc. Inst. Elect. Electronics Engrs.*, **52**, p. 520 [*88*]

FORD, R. (1966): 'The preparation of high resolution silica diffusion barriers by an electron stimulated chemical reaction', Proceedings of the joint IERE–IEE conference on the applications of thin films in electronic engineering [*147*]

FOXELL, C. A. P., and SUMMERS, J. (1966): 'Trends in gallium arsenide varactor diodes' *in* 'Gallium Arsenide', *Inst. Phys. and Phys. Soc. Conf. Ser. 3*, p. 151 [*92, 94*]

FOXELL, C. A. P., and WILSON, K. (1965): 'Gallium arsenide varactor diodes', Proceedings of the joint IERE–IEE symposium on microwave applications of semiconductors, London [*91*]

FROHMAIER, J. H. (1960): 'Noise performance of a three-stage microwave receiver', *Electron. Technol.* **37**, p. 245 [*112*]

FRUIN, A. S.: see ORTON, J. W.

FUKUI, H. (1966): 'The noise performance of microwave transistors', *IEEE Trans.*, **ED-13**, p. 329 [*136*]

GARBRECHT, K. (1965): 'Noise limitations in helium cooled parametric amplifiers', Proceedings of the joint IERE–IEE symposium on microwave applications of semiconductors, London [*94*]

GELNOVATCH, V. (1967): 'The design of distributed transistor amplifiers at microwave frequencies', *Microwave J.*, **10**, p. 41 [*140*]

GINSBERG, T.: see VILCANS, J.

GORDON, J. P., ZEIGER, H. J., and TOWNES, C. H. (1955): 'The maser—new type of microwave amplifier, frequency standard and oscillator', *Phys. Rev.*, **99**, p. 1264 [*36*]

GRANBERRY, D. S., and POLICKY, G. J. (1965): Texas Instruments internal report   [*136*]

GUMMEL, H. K.: see SZE, S. M.

HALL, R. N. (1960): 'Tunnel diodes', *IRE Trans.*, **ED–7**, p. 1   [*118*]

HAMASAKI, H. (1965): 'A low-noise and wide-band Esaki diode amplifier with a comparatively high negative conductance diode at 1·3 GHz', *IEEE Trans.*, **MTT–13**, p. 213   [*124*]

HAMMER, J. M., and THOMAS, E. E. (1965): 'Traveling-wave tube noise figures of 1·0 dB at S-band', *Proc. Inst. Radio Engrs.*, **52**, p. 207   [*16, 20*]

HARRISON, R. I., and ZUCKER, J. (1966): 'Hot carrier microwave detector', *Proc. Inst. Elect. Electronics Engrs.*, **54**, p. 588   [*114*]

HAUS, H. A. (1959): 'Signal and noise propagation along electron beams' *in* SMULLIN, L. D. and HAUS, H. A., (Eds.): 'Noise in electron devices' (Chapman & Hall), chap. 3   [*12*]

HENTLEY, E. L. (1964): 'The construction and performance of a superconducting magnet with high uniform field', Mullard Research Laboratories report 2489   [*55*]

HIELSCHER, F. H.: see SAH, C. T.

HILSDEN, F.: see OXLEY, T.

HOWSON, D. P., OWEN, B., and WRIGHT, G. T. (1965): 'The space-charge varactor', *Solid-State Electronics*, **8**, p. 913   [*90*]

HRBEK, G.: see ADLER, R.

HUGHES, K. L., and PEARSON, J. D. (1966*a*): 'The effect of the upper side band on the performance of a parametric amplifier', *Radio Electronic Engr.*, **32**, p. 337   [*70*]

HUGHES, K. L., and PEARSON, J. D. (1966*b*): 'An S-band parametric amplifier using a balanced idler circuit', *ibid.*, p. 377   [*80*]

HYDE, F. J.: see JONES, W. S.

IRITA, Y.: see MUKAI, H.

ISRAELSEN, B. P., and PETER, R. W. (1962): 'Traveling-wave tube with 1·7 dB noise figure at L-band', *Proc. Inst. Radio Engrs.*, **50**, p. 168   [*20*]

ISRAELSEN, B. P.: see NELSON, J. N.

JAOUN, J. T.: see LORIOU, B.

JELLEY, J. V., and COOPER, B. F. C. (1961): 'An operational ruby maser for observations at 21 cm with a 60 foot radio telescope', *Rev. Sci. Instrum.*, **32**, p. 166   [*43*]

JONES, W. S., and HYDE, F. J. (1967): 'Unilateral 2-diode parametric amplifier', *Proc. IEE*, **114**, p. 1373   [*86*]

JOSENHANS, J. G. (1964): 'Forward bias shot noise in varactor diodes', Bell Telephone Laboratories microwave diode research report 16   [*77*]

159

KERR, J. A., and LARGE, L. N. (1967): 'Semiconductor devices made by ion implantation', Proceedings of the conference on the application of ion beams to semiconductor technology, Grenoble [*147*]

KIKUCHI, C.: see MAHKOV, G.

KLIPHUIS, J. (1961): 'C-band non-degenerate parametric amplifier with 500 MHz bandwidth', *Proc. Inst. Radio Engrs.*, **49**, p. 961 [*80*]

KOSTELNICK, J. J.: see DEGRASSE, R. W.

KOTZEBUE, K. L.: see BLACKWELL, L. A.

KUROKAWA, K. (1965): 'Design theory of balanced transistor amplifiers', *Bell Syst. Tech. J.*, **44**, p. 1675 [*140*]

KUROKAWA, K.: see ENGELBRECHT, R. S.

LAMBE, J.: see MAKHOV, G.

LARGE, L. N.: see KERR, J. A.

LEE, C. W. (1967): 'Push–pull tunnel diode amplifier', Solid-state circuits conference, digest of technical papers, p. 102 [*124, 129*]

LIEGEY, P. R.: see ALLEN, C. M.

LIM, J. T.: see SCANLAN, J. O.

LINVILL, J. C., and GIBBONS, J. F. (1961): 'Transistors and active circuits (McGraw-Hill) [*140*]

LÖCHERER, K. H.: see MAURER, R.

LORIOU, B., and JAOUEN, J. T. (1967): 'C-band traveling wave maser using a slow-wave structure printed on ruby', *Proc. Inst. Elect. Electronics Engrs.*, **55**, p. 461 [*45*]

LUCAS, W. J. (1966): 'Tangential sensitivity of a detector video system with r.f. preamplification', *Proc. IEE*, **113**, p. 1321 [*112*]

LUNT, K. S.: see PEARSON, J. D.

MCCOY, C. T.: see MESSENGER, G. C.

MCPHUN, M. K. (1966): 'A tunable X-band tunnel diode amplifier' *in* 'Microwave and optical generation and amplification'. *IEE Conf. Publ.* 27, p. 292 [*124*]

MCPHUN, M. K. (1967): 'U.H.F. tunnel diode amplifier', *Proc. IEE*, **114**, p. 428 [*119, 124*]

MCWHORTER, A. L., and MEYER, J. W. (1958): 'Solid state maser amplifier', *Phys. Rev.*, **109**, p. 312 [*43*]

MAKHOV, G., KIKUCHI, C., LAMBE, J., and TERHUNE, R. W. (1958): 'Maser action in ruby', *ibid.*, p. 1399 [*43*]

MANLEY, J. M., and ROWE, H. E. (1956): 'Some general properties of non-linear elements. Pt. 1—General energy relations', *Proc. Inst. Radio Engrs.*, **44**, p. 904 [*63*]

MAURER, R., and LÖCHERER, K. H. (1963): 'Low noise non-reciprocal parametric amplifier with power matching at the input and output', *Proc. Inst. Elect. Electronics Engrs.*, **51**, p. 1589 [*86*]

MAVOR, J., and REED, K. B. (1967): 'Equivalent two-port thermal noise representation of MOS transistors', *IEEE Trans.*, **ED-14**, p. 111   [*144*]

MEREDITH, R., and WARNER, F. L. (1963): 'Superheterodyne radiometers for use at 70 GHz and 140 GHz', *ibid.*, **MTT-11**, p. 397   [*87, 103*]

MESSENGER, G. C., and MCCOY, C. T. (1955): 'A low noise figure microwave crystal diode', IRE Convention Record, Pt. 8, p. 68   [*102*]

MEYER, J. W.: see MCWHORTER, A. L.

MEYERER, P. M. (1964): 'New permanent magnet focalizers in high power travelling wave tubes', Proceedings of the 5th international congress on microwave tubes, Paris, p. 206   [*24*]

MILLER, D. J.: see MORRIS, L. C.

MORRIS, L. C., and MILLER, D. J. (1965): 'A C-band rutile traveling-wave maser', *IEEE J. Quantum Electronics*, **QE-1**, p. 164   [*57*]

MUKAI, H., and IRITA, Y. (1967): 'Indium antimonide tunnel diodes', *Rev. Elect. Commun. Lab.*, **15**, p. 37   [*128*]

NELSON, J. N., and ISRAELSEN, B. P. (1965): 'Traveling-wave tube with 2·7 dB noise figure at 12 Gc/s', *Proc. Inst. Elect. Electronics Engrs.*, **53**, p. 548   [*20*]

NIELSON, E. G. (1957): 'Behaviour of noise figure in junction transistors', *Proc. Inst. Radio Engrs.*, **45**, p. 957   [*126, 136, 141*]

NIELSON, E. G. (1960): 'Noise performance of tunnel diodes', *ibid.*, **48**, p. 1903   [*126*]

OKEAN, H. C. (1966a): 'Synthesis of negative resistance reflection amplifiers employing band limited circulators', *IEEE Trans.*, **MTT-14**, p. 323   [*124, 126*]

OKEAN, H. C. (1966b): 'Integrated microwave tunnel diode device', GMTT symposium digest, p. 135   [*124, 130*]

OKEAN, H. C. (1967): 'Microwave amplifiers employing integrated tunnel diode devices', *IEEE Trans.*, **MTT-15**, p. 613   [*132*]

OKWIT, S. (1967): 'Ultra low noise reception using masers and cooled parametric amplifiers' *in* 'Low temperature refrigeration' (Boston Technical Publishers)   [*56, 57*]

OKWIT, S.: see DE GRUYL, J. A.

OKWIT, S., and SMITH, J. G. (1962): 'Packaged electronically tunable S-band traveling wave maser system', *Proc. Inst. Radio Engrs.*, **50**, p. 1470   [*56*]

ORTON, J. W., FRUIN, A. S., and WALLING, J. C. (1966): 'Spin–lattice relaxation of Cr^{3+} single crystals of zinc tungstate', *Proc. Phys. Soc.*, **87**, p. 703   [*48*]

OWEN, B.: see HOWSON, D. P.

OXLEY, T. H., and HILSDEN, F. (1966): 'The performance of backward diodes as mixers and detectors at microwave frequencies', *Radio Electronic Engr.*, **31**, p. 181   [*106, 107*]

OXLEY, T. H., and SUMMERS, J. G. (1966): 'Metal gallium arsenide diodes as mixers' *in* 'Gallium arsenide', *Inst. Phys. and Phys. Soc. Conf. Ser. 3*, p. 138 [*104*]

PAYNE, C. D.: see AITCHISON, C. S.

PEARSON, J. D., and LUNT, K. S. (1964): 'A broadband balanced idler circuit for parametric amplifiers', *Radio Electronic Engr.*, **27**, p. 331 [*80*]

PEARSON, J. D.: see BENNY, A. H.

PEARSON, J. D.: see HUGHES, K. L.

PETER, R. W.: see ISRAELSEN, B. P.

PETER, R. W.: see VEHN, R. E.

PEYTON, B. J.: see ARAMS, F. R.

PIERCE, J. R. (1950): 'Traveling-wave tubes' (van Nostrand) [*10*]

POLICKY, G. J.: see GRANBERRY, D. S.

POMMEREIT, M. (1967): 'Experimental results with a new lower frequency pumped parametric amplifier', *Proc. Inst. Elect. Electronics Engrs.*, **55**, p. 1219 [*85*]

RAUB, W. E., and BERGER, N. W. (1963): 'Experimental results on a low noise high cathode current density traveling wave tube', Electron device research conference, University of Utah [*21*]

REED, K. B.: see MAVOR, J.

ROWE, H. E.: see MANLEY, J. M.

RULISON, R. L.: see FINK, H. J.

SABISKY, E. S., and ANDERSON, C. H. (1967): 'Solid state optically pumped microwave masers', *IEEE J. Quantum Electronics*, **QE-3**, p. 287 [*58*]

SAH, C. T., WU, S. Y., and HIELSCHER, F. H. (1966): 'Effects of fixed bulk charge on the thermal noise in m.o.s. transistors', *IEEE Trans.*, **ED-13**, p. 410 [*144*]

SALZBERG, B.: see ALLEN, C. M.

SCANLAN, J. O. (1966): 'Analysis and synthesis of tunnel diode circuits' (Wiley) [*119*]

SCANLAN, J. O., and LIM, J. T. (1964): 'A design theory for optimum broadband reflection amplifiers', *IEEE Trans.*, **MTT-12**, p. 504 [*124*]

SCANLAN, J. O., and LIM, J. T. (1965): 'The effect of parasitic elements on reflection type tunnel diode amplifier performance', *ibid.*, **MTT-13**, p. 827 [*124, 126*]

SCHULZ-DUBOIS, E. O. (1964): 'Theory of intermodulation and harmonic generation in traveling-wave masers', *Proc. Inst. Elect. Electronics Engrs.*, **52**, p. 644 [*54*]

SCHULZ-DUBOIS, E. O.: see DEGRASSE, R. W.

SCHULZ-DUBOIS, E. O.: see TABOR, W. J.

SCOVIL, H. E. D., FEHER, G., and SEIDEL, H. (1957): 'Operation of a solid state maser', *Phys. Rev.*, **105**, p. 762 [*43*]

SCOVIL, H. E. D.: see DEGRASSE, R. W.

SEIDEL, H.: see SCOVIL, H. E. D.

SEIDEL, H.: see UENOHARA, M.

SMITH, F. W.: see WALLING, J. C.

SMITH, J. G.: see DE GRUYL, J. A.

STEINHOFF, R., and STERZER, F. (1964): 'Microwave tunnel diode amplifiers with large dynamic range'. *RCA Rev.*, **25**, p. 54   [*129*]

STERZER, F.: see STEINHOFF, R.

STEVENS, K. W. H. (1967): 'The theory of paramagnetic relaxation', *Rep. Progr. Phys.* **30**, p. 189   [*48*]

SUMMERS, J. G.: see FOXELL, C. A. P.

SUMMERS, J. G.: see OXLEY, T. H.

SUN, T.: see THORNLEY, R. F. H.

SZE, S. M., and GUMMEL, H. K. (1966): 'Appraisal of the semiconductor–metal–semiconductor transistor', *Solid-State Electronics*, **9**, p. 751   [*146*]

TABOR, W. J., CHEN, F. S., and SCHULZ-DUBOIS, E. O. (1964): 'Measurement of intermodulation and a discussion of dynamic range in a ruby traveling-wave maser', *Proc. Inst. Elect. Electronics Engrs.*, **52**, p. 656   [*52*]

TERHUNE, R. W.: see MAKHOV, G.

TE WINKEL, J. (1959): 'Drift transistor—simplified electrical characteristics', *Radio Electronic Engr.*, **36**, p. 280   [*138*]

THOMAS, E. E.: see HAMMER, J. M.

THOMPSON, G. H. B. (1961): 'Unidirectional lower sideband parametric amplifier without circulators', *Proc. Inst. Radio Engrs.*, **49**, p. 1684   [*86*]

THORNLEY, R. F. H., and SUN, T. (1965): 'Electron beam exposure of photo-resists', *J. Electrochem. Soc.*, **112**, p. 1151   [*147*]

TORREY, H. C., and WHITMER, C. A. (1948): 'Crystal rectifiers' (McGraw-Hill)   [*99, 100, 101, 112*]

TOWNES, C. H.: see GORDON, J. P.

TRAMBARULO, R.: see BURRUS, C. A.

UENOHARA, M. (1967): 'Cooled varactor parametric amplifiers', *Advances in Microwaves*, **2**, p. 89   [*71*]

UENOHARA, M., and ELWARD, J. P. (1964): 'Parametric amplifiers for high sensitivity receivers', *IEEE Trans.*, **MIL–8**, p. 273   [*82*]

UENOHARA, M., and SEIDEL, H. (1961): '961 MHz lower side-band up-convertor for satellite tracking radar', *Bell Syst. Tech. J.*, **40**, p. 1183   [*67*]

VAN DER ZIEL, A.: see BRUNCKE, W. C.

VAN DUZER, T.: see WADHWA, R. P.

VEHN, R. E. and PETER, R. W. (1963): 'Traveling-wave tube with 1·5 dB noise factor in u.h.f. band', *Proc. Inst. Elect. Electronics Engrs.*, **51**, p. 1140 [*20*]

VILCANS, J., and GINSBERG, T. (1964): '35 GHz parametric amplifier with 29 GHz pump', *NEREM Record*, **6**, p. 142 [*85*]

WADHWA, R. P., and VAN DUZER, T. (1965): 'A 3·5 dB noise factor, S-band medium power, forward-wave injected-beam crossed-field amplifier', *Proc. Inst. Elect. Electronics Engrs.*, **53**, p. 425 [*34*]

WALLING, J. C., and SMITH, F. W. (1963): 'Solid state masers and their use in satellite communications systems', *Philips Tech. Rev.*, **25**, p. 289 [*44, 49*]

WALLING, J. C.: see ORTON, J. W.

WARNER, F. L.: see MEREDITH, R.

WATSON, P. A. (1968): 'Shot noise in microwave parametric amplifiers using gallium arsenide varactors', unpublished report [*77*]

WEBER, J. (1953): 'Amplification of microwave radiation by substances not in thermal equilibrium', *IRE Trans.*, **ED-3**, p. 1 [*36*]

WEISS, M. T. (1957): 'A solid state microwave amplifier and oscillator using ferrites', *Phys. Rev.*, **107**, p. 317 [*61*]

WESTCOTT, R. J. (1967): 'Investigation of multiple f.m./f.d.m. carriers through a satellite t.w.t. operating near to saturation', *Proc. IEE*, **114**, p. 726 [*30*]

WHITMER, C. A.: see TORREY, H. C.

WILSON, K.: see FOXELL, C. A. P.

WITTKE, J. P. (1957): 'Molecular amplification and generation of microwaves', *Proc. Inst. Radio Engr.*, **45**, p. 291 [*38*]

WRIGHT, G. T.: see HOWSON, D. P.

WU, S. T.: see SAH, C. T.

YU, H. N. (1963): 'The frequency response of opto-transistors', *Proc. Inst. Elect. Electronics Engrs.*, **51**, p. 945 [*146*]

ZACHARIAS, A. (1961): 'A method for measuring the incremental phase and gain variations of a traveling wave tube', IRE Convention Record Pt. 9, p. 151 [*29*]

ZEIGER, H. J.: see GORDON, J. P.

ZUCKER, J.: see HARRISON, R. I.

INDEX